ALAN DA COSTA GUIMARÃES

EVALUATION OF SEA WATER DESALINATION BY MICROORGANISMS

ALAN DA COSTA GUIMARÃES

EVALUATION OF SEA WATER DESALINATION BY MICROORGANISMS

A RESEARCH ON THE REMOVAL OF DISSOLVED SALTS FROM SEAWATER BY MICROORGANISMS CULTIVATED IN BAMBOO

ScienciaScripts

This book is a translation from the original published under ISBN 978-3-330-75506-2.

Publisher:
Sciencia Scripts
is a trademark of
International Book Market Service Ltd., member of OmniScriptum Publishing Group
17 Meldrum Street, Beau Bassin 71504, Mauritius
Printed at: see last page
ISBN: 978-620-2-97361-8

This work is dedicated to my
family, friends and teachers.

ACKNOWLEDGEMENTS

I thank life for being so spectacular, providing me with so many special and perfect moments.

Gratitude to the divine power, for being so wonderful and present in all moments and places. Force that helps me to feel my intuition to follow the process of being a better human being.

I thank my dear parents, Luiz Carlos Guimarães and Rosa Maria da Costa Silva, for always supporting my choices, trusting me and my work. They are the pillars of my life, I take you wherever I go, your love is inspiring and is my fuel to break through the unknown with confidence, character and true values. All the effort and work that I have perfected and accomplished has been driven by the love and affection that you give me. Father and mother, thank you very much.

My warmest thanks to my brothers, Ighor and Arthur and also to my dear nephew Victor for always supporting me in my projects and encouraging me to do my best. I love you.

I thank my friends of Santos, strong friendships that are kept in my heart, you are part of the construction of the human being that I am today. Thank you Leandro, Marcelo, Pedrinho, Pintado, Vitinho and so many other childhood friends that I take their essences with me.

To my friends at Balneário Camboriú, blessed people who are part of my life, thank you for encouraging me so much to study and to be a good human being. Gratitude, friends.

To my friends from the days of Ireland, Matias, Martin, Fernando, Javier and Bob. The experience on your side has made me great! Thank you for all the support, learning and every moment lived next to you.

I thank my friends from the Sanitary and Ambietal Engineering course at UFSC and from Floripa's life, for so many smiles and laughs, unforgettable walks, exchange of ideas, advice and experiences. I thank the teachers and employees of UFSC for their efforts and providing the best of their work.

I thank the Now, a unique and powerful moment that must be experienced and savoured with presence, peace and tranquillity.

SUMMARY

Desalination is a technique widely used in regions unfavourable to water resources, where salts and minerals dissolved from salt or brackish water are removed. In this sense, the aim of this work was to evaluate different systems of sea water desalination by microorganisms from the inoculum of the bacteria *Pseudomonas stutzeri*. The use of *bamboo*, of the species *Bambusa tuldoides,* as a support medium for the microorganisms in the systems was also evaluated, since the bamboo has starch in its composition, favoring the cultivation of Pseudomonas *stutzeri*. In the present work, nine different reactors were built, varying the concentration of the bacteria, the availability of oxygen, the hydraulic holding time (TDH) and the volume of sea and bamboo water. The water samples were collected in the sea that bathes the coast of Campeche, Florianópolis (SC). The parameters analyzed were: electrical conductivity, pH, apparent colour, turbidity and chlorides. The analyses were performed at the Water Potabilization Laboratory (LAPOA), located at the Sanitary and Environmental Engineering Department of the Federal University of Santa Catarina (UFSC). The reactor that performed best was the one that operated under anaerobic conditions, composed of 6 L of sea water, 500 g of bamboo, with a concentration of 1.67×10^6 forming units of *P. stutzeri* colony per mL of H2Omar, with 62-day TDH. In these conditions the electrical conductivity value decreased by 6.26% (input of 52.7 mS and output of 49.4 mS) and the chloride concentration was reduced by 23.75% (input of 19.99 g/L Cl and output of 15.25 g/L Cl). In general, the systems analysed showed an increase in relation to the apparent colour and turbidity parameters. However, in anaerobic reactors, reductions were presented regarding electrical conductivity and chloride concentration.

Keywords: Bamboo. Desalination. Microorganisms.

SUMMARY

1   INTRODUCTION

Water is the source of life on planet Earth, and this natural resource is essential to most food, energy and manufacturing processes. Planet Earth, also called "Planet Water" has a surface with about 71% of water in a liquid state, of this amount, only 3% is fresh water, where 2.5% is frozen in Antarctica, the Arctic and in glaciers, leaving only 0.5% of fresh water available on the planet to meet the needs of man and ecosystems (ANA, 2009).

World demand for water has increased at a rate of approximately 1% per year as a result of population growth, economic development and changes in consumption patterns, among other factors. An estimated 3.6 billion people (almost half the world's population) live in areas with a potential for water scarcity for at least one month a year (UN, 2018).

Brazil has a large amount of water resources available in the form of rivers, lakes and groundwater, representing about 12% of the planet's fresh water reserves. However, this abundance of this natural resource is irregularly distributed, where approximately 80% is concentrated in the Amazon with a population of only 5% of Brazil's inhabitants. Therefore, 20% of the country's water reserves supply 95% of Brazilians (COUTINHO, 2013). Thus, in order for the population to have access to quality water in the most diverse regions, advances in water treatment processes and technologies are necessary.

Desalination is a technique widely used in regions that are unfavourable for water resources. The desalination process consists in removing dissolved salts and minerals from salt or brackish water, making them in regularity with the standards of potability, defined by the Ministry of Health's Ordinance No. 5/2017.

Technological advances have made it possible to develop various methods for water desalination, such as reverse osmosis, electrodialysis and various techniques based on the distillation principle, already used in several countries such as the United States, the Arab Emirates, Kuwait, Australia, among others (KHAWAJI et al., 2008). The problem with these desalination methods is the consumption of significant amounts of thermal and electrical energy, resulting in high greenhouse gas emissions in the atmosphere (SEMIAT, 2008).

An innovative technique that began to be studied in 2009 is the Microbial Desalination Cells (MDC), a system that is capable of treating wastewater, generating energy and desalinating salt water. There are several studies on variations in the configurations of this system, such as the air cathode CDM (GUDE et al., 2013; SAEED et al., 2015; MEHANNA, 2010), biocathode CDM (SAEED et al., 2015; WALTER et al., 2013) and stack structure CDM (QU, 2013; KIM et al., 2013), among others.

Regarding the use of *Pseudomonas stutzeri* bacteria in the salt removal process, this was used to remove the insoluble salt of potassium nitrate that was encrusted in the paintings of the baroque church San Juanes (Spain), this was the first study analyzing the feasibility of removing insoluble salts using microorganisms in agar, the result was the removal of 92% of this salt, in a short time application of 90 minutes of agar with colonies of *P. stutzeri* (Bosch et al. 2013).

In this sense, this study aims to evaluate the efficiency of different desalination systems: whether or not they contain the inoculum of *Pseudomonas stutzeri*, under aerobic or anaerobic conditions, with or without the presence of bamboo as a support medium.

2   OBJECTIVES

2.1   GENERAL OBJECTIVE

To evaluate the efficiency of different sea water desalination system configurations based on electrical conductivity and chloride concentration, using *Pseudomonas stutzeri* bacteria and bamboo as a support medium for the microorganism.

2.2   SPECIFIC OBJECTIVES

- Analyse the efficiency of the desalination system in the absence or presence of the inoculum of *P. stutzeri* bacteria;
- Analyse the efficiency of the desalination system under aerobic and anaerobic conditions;
- Analyse the efficiency of the desalination system containing or not the bamboo support medium;
- Analyse the efficiency of the desalination system at different hydraulic holding times (TDH);
  Determine the best configuration of the desalination system

3 THEORETICAL BACKGROUND

3.1 CHEMICAL COMPOSITION OF SEA WATER

It is believed that all the planet's natural elements are found dissolved in seawater, although it has not yet been possible to detect some of them. Seawater is simply composed of pure water where various types of solids and gases are dissolved, which can be divided into four classes (SCHMIEGELOW, 2004):

1st Conservative Elements: occur in high concentrations (greater than one milligram per kilo);
2nd Nutrients: essential for the growth of marine plants;
3rd Elements - trace: occur in very small concentrations (less than one milligram per litre);
4th Dissolved Gases.

In sea water solution, 96.5% is pure water and 3.5% is dissolved salts, so 1 kilogram of pure sea water contains 35 grams of dissolved compounds called inorganic salts (SCHMIEGELOW, 2004). Table 1 shows the chemical composition of seawater in order of abundance.

Table 1 - Chemical composition of sea water in order of abundance.

Category	Examples	Concentration range
Conservative elements: Ions more abundant	Cl-, Na+, Mg2+, SO42+, Ca2+, K+	mM
Conservative elements: Less abundant ions	HCO3, Br ⁻, Sr+, F-	μM
Gases	N2, O2, Air, $_{CO2}$, N2O, (CH3)$_{2S}$, H2S, H2, CH4	nM to mM
Nutrients	NO_3^-, NO_2^-, NH_4^+, PO_{444}^{3-} H SiO	μM
Metals (dashes)	Ni, Li, Fe, Mn, Zn, Pb, Cu, Co, U, Hg	< 0.05 μM
Dissolved organic compounds Colloids	Amino acids and humic acids Sea foam and flakes	g.L-1 to mg.L-1
Particulate	Sand, clay, dead fabric, bodies marine, feces	g.L-1 to mg.L-1

Source: Libes (1992).

Preservative elements are the most abundant of the dissolved substances, and although they correspond to 99.28% of the weight of all dissolved material, the most important elements are only 6: Chlorine, Sodium, Sulphate, Magnesium, Calcium and Potassium (SCHMIEGELOW, 2004). Table 1 shows the percentage by weight of the conservative elements present in seawater with salinity of 34.8.

Table 1 - Conservative sea water constituents with salinity of 34.8.

Condoms Elements Percentage by Weight	
Chlorine (Cl-)	55,04
Sodium (Na+)	30,61
Sulphate (SO$_4$ $^{2+}$)	7,68
Magnesium (Mg2+$^)$	3,69
Calcium (Ca2+)	1,16
Potassium (K+)	1,10
Total	99,28

Source: SCHMIEGELOW (2004).

Chlorides are generally distributed in nature as sodium (NaCl), potassium (KCl), and calcium (CaCl2) salts. The largest amount of these salts are present in the oceans (FUNASA, 2006).

Regarding the formation of ionic compounds, the ionic bond of sodium chloride occurs by the reaction between the electropositive element of sodium and the electronegative element of chlorine, since the sodium atom, due to its electronic configuration of 1s2, 2s2, 3s1, presents the first and second complete electronic levels, but the third contains only 1 single electron, therefore, when the sodium atom loses the last electron a stable electronic configuration is obtained (LEE, 1999).

According to Lee (1999), chlorine atoms have an electronic configuration of 1s2, 2s2, 3s2 and 3p5, with only 1 electron missing to reach noble gas conditions, so when the chlorine atoms react, one electron is received and transformed into the chloride ion, with negative charge Cl-.

Sodium chloride is easily obtained by these two types of atoms (sodium and chlorine) to achieve a stable electronic configuration of noble gas, being maintained in a crystalline reticulum through electrostatic attraction (LEE, 1999). The chlorine atom reaches the octet structure by sharing one of its valence electrons with the hydrogen atom, forming the HCl (VOLLHARDT et al., 2013).

All the gases in the atmosphere are dissolved in seawater, and the most abundant gases in the atmosphere are also nitrogen (most abundant of the gases dissolved in seawater), oxygen, argon and carbon dioxide. However, the proportion between them varies, since the solubility of a non-reactive gas depends on three factors: the temperature of the gas and of the solution, partial pressure

gas in the air and liquid and salinity in the case of sea water
(SCHMIEGELOW, 2004).

3.1.1   SALINITY

As regards the origin of the salts found in sea water, cations come mainly from the decomposition of igneous rocks, formed by the crystallization of magma, the result of volcanism. Thus, the physical-chemical decomposition of these rocks results in fragments of smaller pieces, releasing ions that are carried to the sea through rivers (SCHMIEGELOW, 2004).

Anions are present in the interior of the planet and may also have been present in the primitive atmosphere, these negatively charged ions are released as gases in the occurrence of volcanic eruptions, move into the atmosphere and return as rainwater precipitation, in the occurrence of marine volcanic eruptions, the anions can be dissolved directly in sea water (SCHMIEGELOW, 2004).

Salinity can be defined as the total concentration of dissolved ions in water (LOGAN, 1965). Thus, salinity, total dissolved solids (STD) and electrical conductivity are directly proportional since ions are electrically charged particles (OLIVEIRA, 2005).

Salt is broken down into ions when dissolved in water, being cations when positive and anions when negative (SCHMIEGELOW, 2004). Dissolved substances include inorganic salts, organic substances and gases and in general, salinity shows little variation between different oceans and seas: the average salinity in the Pacific is 35.5 ‰, in the Atlantic 37.0 ‰ and in the Red Sea 40.0 ‰ (RESENDE, 2009).

According to Silva (2011), salinity is considered a conservative property because its concentration does not change due to biochemical processes, unlike dissolved oxygen (O_2) and carbon dioxide (CO_2). Thus, dilution (river discharge or precipitation) and evaporation are the processes responsible for the variation in salinity.

The amount of salts dissolved in sea water is around 35 ‰, much higher than the concentration found in river water, where it does not exceed 0.2 ‰ of dissolved salts (SILVA, 2011).

From Table2, which compares the concentration of the chemical composition of river and seawater, the dominance of calcium bicaborate (HCO $^-$) in fresh water can be seen, while chloride and sodium represent the most inorganic components found in seawater (SILVA, 2011). Table 2 also shows that a higher concentration variation of inorganic elements occurs in freshwater than in seawater.

Table 2 - Composition of the main inorganic components found in the waters of rivers that flow into the main oceans.

Components	Composition of the Waters of the Rivers which Land in the Oceans mg/L	Average of Rivers mg/L	Sea Water mg/L

	Atlantic	Pacific	Indian		
Cl-5	,7	5,1	6,8	5,9	19.344
Na+	4,2	5,2	8,5	6,0	10.733
SO47	,7	9,2	7,9	8,3	2.712
Mg2+2	,5	3,6	5,4	3,8	1.294
Ca2+10	,5	13,9	21,6	15,3	412
K+1	,4	1,2	2,5	1,7	399
HCO3	37	55,4	94,9	62,4	142
SiO29	,9	11,7	14,7	12,1	-

Source: SILVA (2011).

The horizontal distribution of salinity in the oceans varies according to latitude, the salinity is minimal near the equator, a region where the atmospheric pressure is low, therefore the warm air rises causing clouds and rainfall. Thus, the high rainfall added to relatively less intense evaporation than in tropical regions results in low salinity in this region (PEREIRA et al., 2009).

In the tropics, salinity is higher as there is great surface evaporation caused by trade winds, constant winds that are common in these regions, and also, the atmospheric pressure in tropical regions is high, which causes relatively small rainfall, already in the poles, the thawing makes the water less salty (PEREIRA et al., 2009).

As for the vertical distribution, at medium and low latitudes, the salinity is higher at the surface and lower towards the bottom and this occurs due to evaporation, the inverse occurs at high latitudes due to thawing, since fresh water remains at the surface because it is less dense (SCHMIEGELOW, 2004). Despite the variations in salinity, the relationships between the most abundant ions remain the same (PEREIRA et al. 2009).

Regarding the hydrogenation potential (pH) of seawater, the presence of CO_2 and highly alkaline ions such as sodium, potassium and calcium ions tend to make seawater slightly alkaline with values between 7.5 and 8.4 (RESENDE, 2009).

3.2 BIOLOGICAL COMPOSITION OF SEAWATER Marine

microorganisms are all organisms that are smaller than 100 µm, including: cyanobacteria, heterotrophic bacteria, archaeas, eukaryotic phytoplankton and a diverse array of autotrophic, heterotrophic and myxotrophic protista and also viruses, the most numerous entities in the oceans. These microorganisms are essential in the mediation of various biochemical cycles, as well as other processes that control greenhouse gases (KIRCHMAN et al., 2000).

The aerobic and anaerobic biogeochemical cycles play an important role in the removal of organic matter, and in the aerobic cycle microorganisms use oxygen as the final electron acceptor to perform the decomposition of organic matter, already in anaerobic, oxygen is not present, so it is the metabolism that promotes the rearrangement of the electrons of the molecule, the anaerobic digestion is characterized by the production of methane and carbon dioxide (BUSWELL & MUELLER, 1952).

In the study by Belila and collaborators (2016) on the structure and variation of the bacterial community in seawater in a large-scale desalination plant for the production of drinking water, several bacterial philosophies were detected in the samples. Proteobacteria phylum bacteria (48.6%) had the highest domain, followed by Bacteroidetes (15%), Firmicutes (9.3%), Cyanobacteria (4.9%),
Planctomycetes (3.0%) and Actinibacteria (1.9%).

For Proteobacteria, 6 classes were identified in all sample processes: Alphaproteobacteria (29.5%), Gammaproteobacteria (2.3%) (Pseudomonadales account for 24.5% to 65% of this class composition), Deltaproteobacteria (2.7%) and Betaproteobacteria (2.3%) (BELILA et al., 2016)

3.2.1 PSEUDOMONAS STUTZERI

According to Lalucat and collaborators (2006), *Pseudomonas stutzeri* are members of the genus Pseudomonas sensu stricto, belong to the phylum Proteobacteria and class Gammaproteobacteria.

Pseudomonas stutzeri, is an optional autotrophic species, capable of performing autotrophic oxidation of sulphur compounds coupled with nitrogen compound reduction (MAHMOOD et al., 2009).

Lalucat and collaborators (2006) explain that *Pseudomonas stutzeri* is an ubiquitous bacterium with a high degree of physiological and genetic adaptability, present in a large number of different natural environments (soil, rhizosphere, groundwater, oceans, sediments, salt marshes and sewage treatment plant effluent) and like other Pseudomonas species, is involved in important metabolic activities such as metal cycling and degradation of biogenic and xenobiotic compounds (petroleum derivatives - aromatic and non-aromatic hydrocarbons - and biocides).

Phenotypic categories of the genus *Pseudomonas stutzeri* include, Gram negative staining, positive catalase tests (enzyme that breaks down hydrogen peroxide (H_2O_2) molecules into water and oxygen) and oxidase (enzyme that catalyzes oxidation and reduction reactions involving molecular oxygen as electron acceptor). They have strictly respiratory metabolism, with strains defined as denitrifying, which can grow in starch and maltose and have negative reaction in hydrolysis tests of arginine dihydrolase and glycogen. They are usually stem-shaped cells 1 to 3 μm long and 0.5 μm wide and single polar flagellum (LALUCAT et al., 2006).

P. stutzeri can grow in a wide range of temperatures, ranging from 4° to 45° C, where most strains grow between 40 and 43°C, but the optimum temperature for growth is approximately 35°C. These microorganisms do not tolerate acid conditions, the growth being inhibited at pH 4.5 and they grow well under atmospheric oxygen and can be facultatively anaerobic with nitrate and in some strains they can also be anaerobic with chlorate or perchlorate as electron acceptors. For a strain to be considered of marine origin, its physiological characteristics must require or at least tolerate the presence of NaCl in the medium. Thus, *P. stutzeri, P. balearica and P. xanthomarina* appear to be marine

Pseudomonas species (LALUCAT et al., 2006).

Like most known *Pseudomonas*, *P. stutzeri* strains can grow in chemically defined media, with ammonia or nitrate ions and a single organic molecule as the sole source of carbon and energy, requiring no additional growth factor. Some rare strains of this microorganism can grow diazotrophically (by fixing atmospheric nitrogen). In general, they do not tolerate acid conditions (they do not grow at pH 4.5), they have respiratory metabolism and oxygen is the terminal electron receptor, nitrate can be the alternative electron acceptor and can perform denitrification (LALUCAT, 2006).

In relation to structures and types of colonies, *P. stutzeri* can be distinguished by the unusual shape and consistency of its colonies (Figure 1), with newly isolated colonies being adherent, with a characteristic wrinkled appearance with a reddish brown colour, typically hard and dry, easily removed from a solid surface (LALUCAT, 2006).

Figure 1- Various types of morphology of *P. stutzeri* colonies.

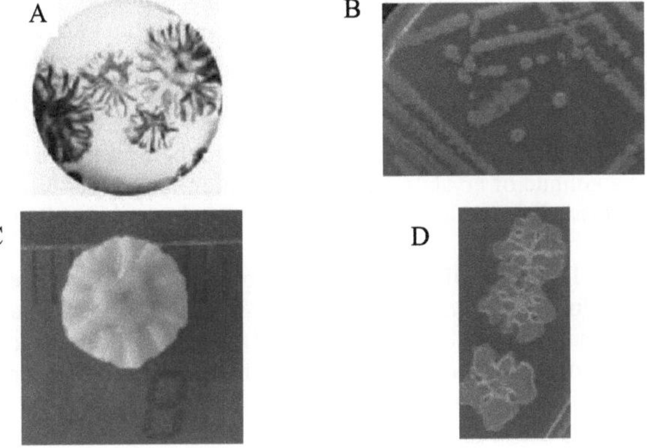

Source: LALUCAT (2006).

3.2.1.1 REMOVAL OF UNDISSOLVED SALTS BY MEANS OF
THE PSEUDOMONAS STUTZERI BACTERIA

The bacteria *Pseudomonas stutzeri* is a nitrate reducer and was used in the study by Bosch and collaborators (2010) for the first time to remove the nitrate salt crust from the artistic surfaces of the Baroque church Santos Juanes (1240), located in Valencia, Spain. The bacteria *Pseudomonas stutzeri* converts nitrates into molecular nitrogen, a gas that evaporates into the atmosphere at room temperature (Ranalli and Sorlini 2003).

The wall paintings were made by the fresco technique where the pigments are applied with water and fixed to the lime and sand mortar by carbonation processes, in this case the formation of saline efflorescence in this painting is due to an internal space present at the bottom of these murals, where multiple rodents, birds and insects lived, generating large accumulations of organic waste and producing nitrate from the oxidation of nitrogen present in organic matter, which with the help of rain leaks through the wall and forms a white efflorescence on the surface of the wall painting (Bosch et al., 2013).

The main chemical composition of the saline efflorescence found in the mural painting of Santos Juanes was potassium nitrate (Bosh et al., 2013). Domeneche and Yusa (2006) explain that the formation of saline efflorescence on the surface of paintings is one of the most important mechanisms of deterioration of works of art in indoor environments since the precipitation of salt exerts a pressure on the wall due to the increase in the volume of crystals, generating traction forces that can exceed the resistance of the material causing micro-cracks in the wall and layer of the painting.

The work of Bosch and collaborators (2013) showed that the application of *P. stutzeri* on agar for a short time of 90 minutes resulted in a good cleaning of the insoluble efflorescent nitrate deposit on the surfaces of wall paintings, the analysis of ion chromatography proved that there was a 92% reduction in efflorescent nitrate.

3.2.2   ARCHAEABACTERIA

The Archaea domain is composed of a group of bacteria with extremophilic characteristics. These micro-organisms tolerate extreme environmental conditions such as very saline environments (halophilic), very low pH (acidophilic) and very high temperatures (hyperthermophilic) (SPOLIDORIO et al, 2015).

Studies show that archaea is found in abundance in deep oceans, are chemilithoautotrophic (microorganism capable of producing its own food substances from the energy released by chemical reactions between inorganic components of the earth's crust). Thus, these micro-organisms oxidize ammonia (chemilitotrophy) to conduct the fixation of CO_2 (chemoautotrophy) (KIRCHMAN et al., 2000). Ingalls and collaborators (2006) suggest that 80% of the carbon used by archaeas comes from CO_2 assimilated via chemoautotrophy at a depth of 670 metres in the North Pacific.

Archaeabacteria live in environments with a high concentration of salt that could kill most microorganisms. According to Arora (2014), salt in the strategy is employed by real halophiles, including archaea halophilica and extremely halophilic bacteria. These micro-organisms are adapted to the high salt concentrations and cannot survive when the salinity of the medium is lowered.

These microorganisms generally do not synthesize organic solutes to maintain the osmotic balance. The intracellular $K+$ concentration is generally higher than the external, while the intracellular $Na+$ concentration is generally lower than in the middle, the intracellular $K+$ concentration increases with the increase of the external NaCl concentration in a non-linear pattern. All halophilic microorganisms contain powerful transport mechanisms, usually based on $Na+$ / $H+$ (OREN, 1999).

Halophilic bacteria have developed mechanisms in the cell membrane to survive in hostile environments, known as osmoadaptation, these microorganisms accumulate organic ions ($K+$, $Na+$ and $Cl-$) in high concentrations in such a way as to balance the osmotic pressure inside with that of the external environment, and thus make it possible to preserve their cellular integrity (PILLING, 2015).

In *cold* springs (*cold seeps*) there is a type of ecosystem supported by chemosynthesis practiced by bacteria and archaeas, where substances such as methane (and other hydrocarbons) and sulphide flow through the substrate and are used as raw material for the synthesis of organic matter by chemosynthetic prokaryotes. This ecosystem has special chemosynthetic bacteria, methanotrostrophic bacteria, in addition to those that oxidize sulfur salts and other minerals (PEREIRA, 2009).

Little is known about marine archaeas, chemilithoautotrophy and the various microbes and processes of the ocean

deep, since most studies are focused on the shallowest layer of the oceans and on euphotic zones (part of the aquatic ecosystem that receives enough sunlight for primary production). In the water mass below the euphtic zone, known as the dark ocean, not many samples occur yet, especially because of its vastness, since 75% of the ocean is below 1000 m, which makes the dark ocean the largest habitat in the biosphere (KIRCHMAN et al., 2000).

3.3 BACTERIAL METABOLISM

Metabolism is the set of interconnected biochemical reactions of a living being, it is a very complex network of chemical reactions, since there are a large number of intermediate compounds present in the processes that allow the maintenance and duplication of a cell (FACINCANI, 2002).

The metabolism is made up of a series of reactions in which the great challenge is to obtain ATP (adenosine triphosphate), NADPH (nicotinamide adenine dinucleotide phosphate) and 13 precursor molecules which will be used in the process of biosynthesis of all cell constituents, as well as in cell maintenance (FACINCANI, 2002).

In glycolysis, one glucose molecule (six-carbon compound) is converted into fructose (six-carbon compound), which eventually forms two pyruvate molecules or pyruvic acid (three-carbon compound, C3H4O3). The glycolytic pathway (Embden-Meyerhoff pathway), involves many steps, including the reactions in which the metabolics of glycolysis are oxidized.

Pyruvate has many destinations when formed, according to Campbell, (2000) the formation of pyruvate (C3H4O3) or pyruvic acid originates at the end of glycolysis and its amount of pyruvate is limited in the cell, so it needs to be regenerated and can be made from the reduction of pyruvate in:

1° - Ethyl acid (ethanol) from alcoholic fermentation; 2° - Lactic acid (lactate) from anaerobic glycolysis;

3° - Carbon dioxide, water and acetyl - CoA (citric acid cycle) from aerobic oxidation.

Acids form ionic solutions, i.e. they generate free ions, guaranteeing the conductivity of electric current (FACINCANI, 2002).

According to Facincani, (2002) the Entner - Doudoroff route is described only in microorganisms and is an alternative for glucose metabolism. This is the main route of glucose degradation in several Gram-negative bacteria: *Pseudomonsa, Zymomonas,*

Ralstonia, Xanthomonas etc. The Via Entner - Doudoroff is used for the degradation of other substrates and is present in microorganisms that live in environments with glyconate availability. The net result of the degradation of glucose by this route is represented by the following chemical equation:

$$Glucose + ADP + NADP+ + NAD+ + Pi \rightarrow 2\ pyruvate + ATP + NADPH + NADH \tag{1}$$

Pseudomonas stutzeri in anaerobiosis, does not ferment, therefore presents only anaerobic respiration and the Krebs cycle is complete, NADH being the initial electron donor of the anaerobic electron transport chain (FACINCANI, 2002).

Regarding the chemical composition of bacteria, all microbial nutrients originate from chemical elements, cells consist mainly of macromolecules such as proteins, nucleic acids, lipids and polysaccharides, and in general, carbon and nitrogen are macronutrients in any microbial cell (MADIGAN, 2016). Table 3 shows the percentage of bacteria element composition by dry weight.

Table 3 - Percentage of bacteria element composition by dry weight.

ElementBacteria	
Carbono50	,0 - 53,0
Hydrogen7	,0
Nitrogen12	.0 - 15.0
Phosphorus2	,0 - 3,0
Sulphur 0	,2 - 1,0
Potassium1	,0 - 4,5
Sodium 0	,5 - 1,0
Calcium 0	,01 - 1,1
Magnesium0	,1 - 0,5
Chloride 0	,5
Ferro0	,02 - 0,2

Source: Luria, 1960; Herbert, 1976; Aiba et al., 1973.

3.4 BAMBU (Bambusa tuldoides)

According to Liese (1998), bamboo belongs to the grass family (Poaceae), subfamily bambusoideae and class

Monocotyledoneae, Angiospermae division, there are approximately 75 genera and over 1200 species registered.

The stem of the bamboo is called thatched, composed of us and internodes. The thatch of bamboo is formed by fibro-vascular bundles (60 to 70% of its mass) and parenchymatous tissue rich in starch. The presence of starch is the main characteristic of bamboo stalks, and besides starch there are other dosed components like holocellulose (~65%) and lignin (~18%) (AZZINI and GONDIM-TOMAZ, 1996). Table 4 shows the chemical composition of bamboo.

Table 4: Chemical composition of bamboo.

Component	% by mass
Carbon	50,00
Hydrogen	6,10
Oxygen	43,00
Nitrogen	0,04 – 0,26
Ashes	0,20 – 0,60

Source: BARBOSA (1999).

According to Rozman (1988), lignocellulosic materials have polar hydroxyl groups due predominantly to cellulose and lignin, and these groups are easy to interact with polar polymer matrices, as is the case with polymer resins. Starch is a polysaccharide with the formula $(C6H10O5)_n$, this carbohydrate is mainly composed of glucose with glycosidic bonds, is formed as a product of the cellular activity of chlorophyllate plants and serves as a food reserve (ALMEIDA, 2009).

Parenchyma cells in knots and/or diaphragms have a high starch content, and fibres and protoxylem may also have starch inclusions (LIESE, 1998). Starch provides energy for wall thickening and is attractive to xylophagous organisms, the base of the thatch contains less starch, and in the cold months higher starch contents are observed as a way of meeting demand in the growing season (SULTHONI, 1987). The presence of starch granules is observed in a pot element, as shown in Figure2.

Figure 2 - Cross-section of the stem parenchyma of the giant Dendrocalamus with 3,000 X increase with starch grains.

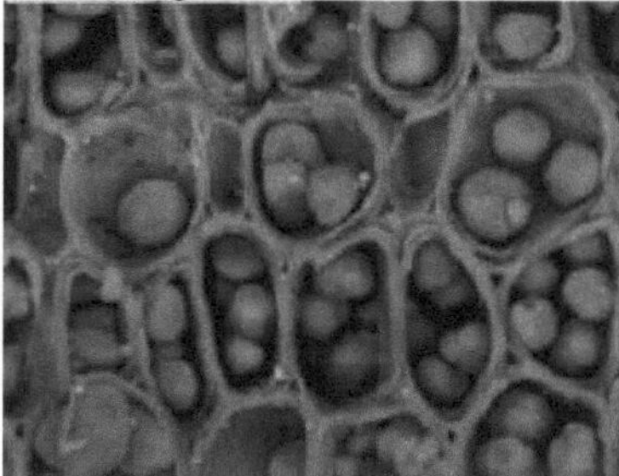

Source - Mirror, J.C.C., (2007).

In relation to the age of bamboo, mature stalks range from 3 to 5 years, and the lignification process is already closed and mechanical property has higher values. Azzini (1987), studied the distribution of starch in the axial and radial directions of the *Bambusa thatch Vulgaris Schrad.* and concluded that, in the radial direction, the starch content predominated in the internal region of the thatch wall (35.1%) and decreased towards the external wall (20.0%), already in the axial direction, there were no significant variations in the starch contents (26.2 to 29.7%).

Although there are no published studies on the application of bamboo in desalination systems, bamboo is already used in natural effluent treatment systems, as in the one developed by Limas (2009), who developed a biological effluent treatment system, using bamboo as a support medium, the system removes pollutants such as nitrate (NO3-) and nitrite (NO2-) from sewage.

The configuration of this effluent treatment system consists of a pre-filtration box followed by a series of 3 septic tank type filled with *bamboo, bambusa vulgaris* and *bambusa tuldoides*, and a filter in the outlet tank, composed internally of layers of coal, bamboo fibres and gravel, connected to each other by tubes, as can be seen in Figure 3 (LIMAS, 2009).

Figure 3 - Natural effluent treatment system using bamboo as a support medium.

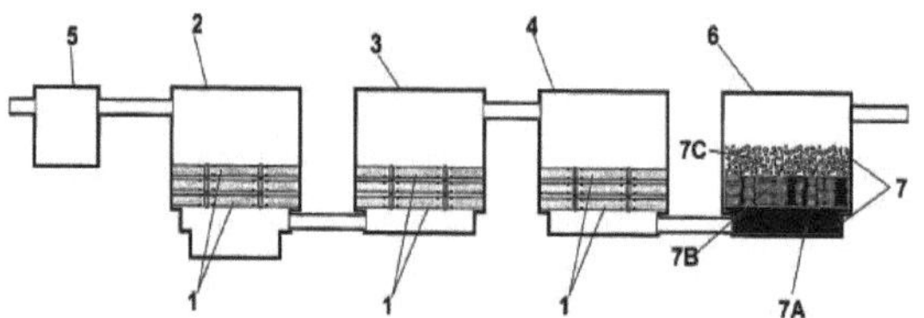

Source: LIMAS (2009).
Note - 1 = half bamboo support, 2, 3 and 4 = modular series tanks, internally fitted with bamboo filter elements, 5 = pre-filtration box, 6 = output tank, 7 = composite output filter composed of carbon (7A), bamboo fibres (7B) and gravel (7C).

Bamboo is used as raw material for feeding and hosting bacteria of the Pseudomonas *liminuis*, Bacilius *phagiens* and Spirillum *brasiliensis* genera where bacteria of the Pseudomonas and Bacilius genera use nitrates as hydrogen carriers in their anaerobic respiration to carry out the denitrification process, transforming nitrates (NO_3^-) sequentially into nitrites (NO_2^-), nitrous oxide (N_2O) and nitrogen gas (N_2) (LIMAS, 2009).

According to Limas (2009), the bacteria present in bamboo They also act in an aerobic environment, since they are an optional aerobic bacteria variety, and can act both in the presence and absence of oxygen without altering its efficiency.

The volume of bamboos is preferably between 7 and 23% of the volume of each filtering tank, and to accelerate the growth of bacteria colonies, 0.02% to 0.07% sodium acetate or potassium acetate and 0.02% to 1% sugar is added to the first tank. In just 2 to 5 hours after these additions are applied, the acceleration of bacteria proliferation is promoted on the internal and external walls of the bamboos and can reach levels that depollute the water on the first day of use (LIMAS, 2009).

In the natural process, i.e. without the use of bacterial growth accelerator additives, satisfactory efficiency in removing pollutants can take on average 120 days. This is why bamboo is not as efficient when used as a filtering element in the first days, since the amount of bacteria

24

is not sufficient for the consumption of pollutants. Bamboo expels a resin in the first days of use, increasing the amount of colour in the water, this process takes on average 90 days to cease and the water shows acceptable colouring (LIMAS, 2009).

According to Limas (2009), the shapes and sizes of bamboos can be in buds, where holes are made in the sides for the inlet and outlet of water, bamboos can also be used in other forms such as discs, chopped, fibers, among others.

Muller (2016) studied the use of bamboo as a filter medium for water treatment and noted that as the length of bamboo increases, so does electrical conductivity, since electrical conductivity is one of the parameters that can indicate the passage of substances from bamboo to the water produced. Therefore, as the length of the bamboo increases, it is observed that these substances are more carried into the water produced.

3.5 BIOFILM

Biofilm is the accumulation of bacteria in a microbial community with complex layers of micro colonies, which can accommodate one or more microbial species to live together by adhering to the polymer matrix produced by them (SAEED et al., 2015). Biofilms are bound by a matrix of polysaccharides and proteins known as sludge (OTTO, 2004).

Colony formation is restricted by the suitability of the surface and surrounding environmental conditions including temperature and humidity level (LÓPEZ, 2010).

Once they have found the ideal surface conditions, the cells adhere and begin to multiply by cell division. In this way, biofilm allows the arrival of other microorganisms at the surface, as it provides diverse sites of adhesion (SAEED et al., 2015).

The transfer of biological material between species is an asset to the environment as it helps to provide shelter for the whole community (HARRISON et al., 2005). Sekar (2002) cites that the thickness, biomass and density of biofilm are directly proportional to their age.

According to Jacovide and collaborators (2010) the growth of biofilm is a slow process, and there are four stages in the life cycle of the biofilm: the 1st stage is separated into stage 1a and 1b being 1a the planktonic cell stage, stage 1b represents the adhesion of the cells in the surface and the formation of a monolayer with united cells; phase 2b is the adhesion between cells and their proliferation; phase 3 is the maturation and 4 the detachment (Figure 4). The fourth stage, "detachment", can leave behind a hollow hill that leads to the completion of the biofilm (PROAL, 2008).

Figure 4 - Stages of biofilm formation on surfaces.

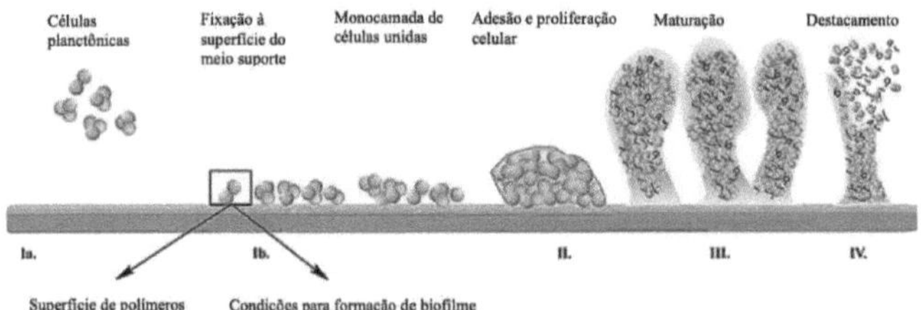

Source: Adapted from EL - DESSOUKY et al., 1995

3.6 METHODS OF DESSALINIZATION OF SEAWATER

According to Shannon and collaborators (2010) the only

methods
capable of increasing water availability beyond what is supplied
by the hydrological cycle, is through desalination and the reuse of water.

Technological advances have made it possible to develop various methods for desalination of sea water. Among the main processes used are reverse osmosis (RO), multi-stage flash distillation (MEF), multi-effect distillation (DME) and electrodialysis (ED). Desalination has been used to obtain drinking water in several countries around the globe such as the United States, Arab Emirates, Kuwait, Australia, among others (KHAWAJI et al., 2008).

Many of the modern large-scale desalination plants are located in the arid countries of the Arab Gulf, most of which operate by thermal processes, in which water is heated followed by evaporation and condensation to produce drinking water (NATIONAL RESEARCH

26

COUNCIL et al., 2008). The problem with these desalination methods is the consumption of significant amounts of thermal and electrical energy, resulting in high greenhouse gas emissions in the atmosphere (SEMIAT, 2008).

The energy required to operate a Reverse Osmosis desalination plant has declined dramatically over the last 40 years (FRITZMANN et al., 2007; BUSCH et al., 2004; ELIMELECH et al, 2011), this decrease is attributed to continuous improvements in desalination technologies, including high permeability membranes, the installation of energy recovery devices and the use of increasingly efficient pumps (FRITZMANN et al., 2007).

Reverse Osmosis desalination plants consume between 3 to 4 kWh/m3 and emit between 1.4 to 1.8 kg of CO_2 per cubic meter of water produced (FRITZMANN et al., 2007; VON et al., 2005; RALUY et al., 2006). In order to understand the size of the projects and their impacts on the environment, Spain would require 400 GWh/year to operate the desalination plant designed to supply 1 billion m3/year (VON et al., 2005).

The two main processes used to reduce the concentration of dissolved salts in brackish water and seawater are thermal processes and membrane processes, thermal processes are based on distillation, where the salt or brackish water is heated so that evaporation occurs, the pure water is then obtained from the condensation of the steam generated (Foundation For Water Research, 2011).

For membrane desalination processes, electrodialysis uses a battery system with ion exchange membranes, which selects both positive and negative ions (Foundation For Water Research, 2011). Thus when, for example, sodium chloride dissolved in water results in positively charged sodium ions and negatively charged chlorine ions, which when under the influence of electric current, the sodium ions pass through the cationic membrane and the negative chlorine ions pass through the anionic membranes (Foundation For Water Research, 2011).

The osmosis process consists of the passage of water through a semi-permeable membrane from a low concentration solution to a high concentration solution, the osmotic pressure is the pressure required for the process to occur (Foundation For Water Research, 2011). Osmosis is a process that occurs in plant and animal tissues, including the human body, however, if the pressure is applied on the side of the membrane with the highest concentration, the reverse osmosis process occurs, where the water with high concentration passes through the semipermeable membrane into a solution with low concentration, for this to occur, the pressure applied must be higher than the osmotic pressure of the system (Foundation For Water Research, 2011).

3.6.1   DESALINATION BY CELLS   OF MICROBIAL DESALINATION (CDM)

The microbial desalination cell (MDC) was first studied in 2009 by a group of researchers at Tsinghua University in China in collaboration with Pennsylvania State University (MEHANNA et al, 2010). Other desalination techniques may require 6 to 68 kWh to desalinate 1 ^{m3} of seawater, while CDMs can produce 180 to 231% more energy in the form of H2 for the desalination of 30 to 5 g/L sodium chloride solutions (WANG et al, 2013). According to Qu (2013) CDMs under certain conditions can produce energy in the form of electricity or hydrogen gas from different sources of renewable organic matter.

CDM is an extension of the microbial fuel cell (CCM). CCM is a technology composed of an anode, cathode, selective cation membrane and an outer wire, as shown in Figure 5.

Figure 5 - General scheme of a microbial fuel cell.

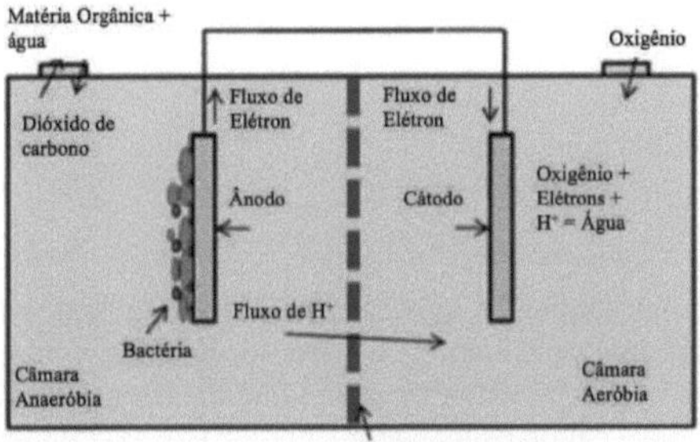

Source: Adapted from SAEED et al., 2015.

Anaerobic conditions are maintained at the anode and aerobic conditions at the cathode. MCCs can operate with or without a mediator, mediated fuel cells involve the external addition of bacteria to oxidize the substrate. While MCCs without mediators do not require the addition of bacteria, they use the bacteria present in the sludge that can be electrochemically activated (SAEED et al., 2015).

In relation to electricity production from wastewater, wastewater containing organic matter enters the anodic side, where available bacteria are proliferated and form a thick cellular aggregate known as biofilm (TORRES, 2012).

The biofilm adheres to the anode and the biocatalysis process is initiated, where the bacteria oxidizes organic matter releasing protons and electrons. In this way the electrons move from the anode to the cathode through an external wire connecting the two electrodes, where the electrode can be exposed to air or surrounded by water in aerobic conditions, then the bioelectricity is produced due to the difference in potential between the cathode chamber and the anode, leading to desalination to occur (SAEED et al., 2015).

According to Saeed (2015), protons pass to the cathode through the selective cation membrane, and then are combined with oxygen and electrons from the external circuit to form pure water. The following equations show the reactions at the anode and cathode:

At the anode:

$$(CH_2O)_n + nH_2O \rightarrow nCO_2 + 4ne^- + 4nH^+ \tag{2}$$

In the cathode:

$$O_2 + 4ne^- + 4nH^+ \rightarrow 2H_2O \tag{3}$$

The CDM unit typically consists of the anode and cathode chamber and an additional desalination chamber located in the middle of the unit, constructed by the insertion of an anionic exchange membrane (ATM) on one side and a cation exchange membrane (TCM) on the other side (Figure 6).

Figure 6 - Typical scheme of a microbial desalination cell (MDC) containing the anode chamber, cathode chamber, desalination chamber located in the middle of the unit and anionic exchange membranes (ATM) and cation exchange membranes (TCM) located on opposite sides.

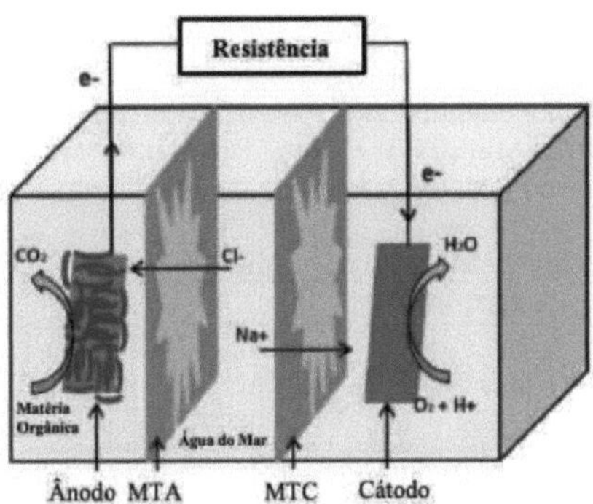

Source: Adapted from PING et al., 2013.

In this way, the anode chamber is responsible for the degradation of organic matter and the production of electricity, the middle chamber is responsible for the removal of salt from sea water while the cathode chamber completes the electrical cycle (LUO et al., 2012).

At the anode the bacteria oxidize organic matter to $_{CO2}$ and H+, released in the anolite, the electrons flow to the cathode through an external electrical circuit, and a current along the cell is established, the receptors of external electrons in the cathode chamber, usually O2, use these electrons to reduce and produce water, This causes a potential gradient through the anode and cathode chamber and to maintain electro-neutrality, anions (Cl- and SO42-) migrate from the salt water in the intermediate chamber through the MTA to the anode, while cations (such as Na+ and Ca2+) move through the MTC to the cathode chamber (SAEED et al., 2015).

According to Forrestal and collaborators (2012), this process can remove more than 99% salt from salt water and can at the same time generate more energy than the external energy required to operate the system. According to Saeed and collaborators (2015), there are different configurations of CDMs: Air cathode CDM, Bio cathode CDM, Stacking frame CDM, Recirculation CDM, Microbial Electrolysis and Chemical

31

Production Desalination Cell (MCPDE), Capacitive CDM, Ascending flow CDM, Osmotic CDM, Bipolar membrane CDM, Decoupled CDM, Stacked circulation separator coupled CDM, Ion exchange resin coupled CDM.

3.7   LEGISLATION

The Ministry of Health's Consolidation Ordinance No. 5 of September 28, 2017 deals with the consolidation of the rules on health actions and services of the Single Health System. The organoleptic standards of potability that pose a health risk, defined by Order No. 5/2017 of the Ministry of Health (Table 5), will be used for the purpose of comparison with the results obtained in this study.

Table 5 - Organoleptic Standard of Potability.

Parameter	VMP	Unit[1]
Chloride	mg/L	250
Apparent Color	uH[2]	15
Turbidity	UT[3]	5

Source: Ministry of Health Ordinance No. 5/2017
NOTES: (1) Maximum Allowable Value; (2) Hazen Unit (mgPt-Co/L); (3) Turbidity Unit.

The organoleptic pattern consists of the set of parameters characterized by causing sensory stimuli that affect acceptance for human consumption, but do not necessarily imply health risk (Ministry of Health, Ordinance 05/2017).

Colour is characterised by apparent colour and true colour, in which the suspended particles are considered, unlike true colour, which is determined after filtering the sample, the apparent colour VMP is 15 uH (Table 5)(Ministry of Health, Ordinance 05/2017).

Turbidity is defined as a measure of the degree of interference with the passage of light through the liquid, the change in the light in the water derives from materials in suspension, this parameter is expressed by means of turbidity units, the VMP for turbidity is 5 UT (Table 5) (Ministry of Health, Ordinance 05/2017).

According to the standards set by the Ministry of Health, the chloride concentration should not exceed 250 mg/L Cl (Table 5).

Ordinance 05/2017 does not specify the maximum permitted

values for electrical conductivity, therefore the Portuguese Decree-Law 306/ 2007 will be used as a reference, where the value for electrical conductivity established for the control of the quality of water intended for human consumption is 2,500 µS/cm at 20°.

4 MATERIALS AND METHODS

The present work was carried out in two stages, in the 1st stage the isolation, identification and culture of the *Pseudomonas stutzeri* bacterium (Figure 7), this stage is detailed in item 4.2.

Figure 7 - 1st Step of the methodology - Isolation, identification and cultivation of *P. stutzeri*.

The second stage of the work is the operation of the reactors.

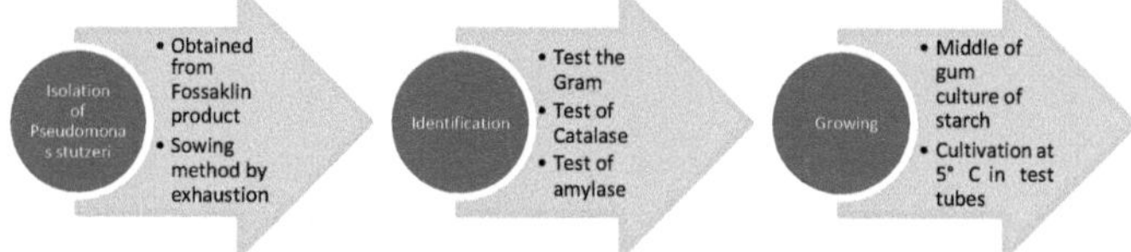

Nine different reactors were built, varying in bacterial concentration, oxygen availability, hydraulic holding time, and sea and bamboo water volume, the details of each reactor are presented in item 4.5.

The stages of the pre-operation consist of collecting sea water, harvesting the bamboo, building the reactors, *P. stutzeri's inoculum* and starting the operations, as summarised in the diagram in Figure8.

Figure 8 - 2° Stages of pre-operation.

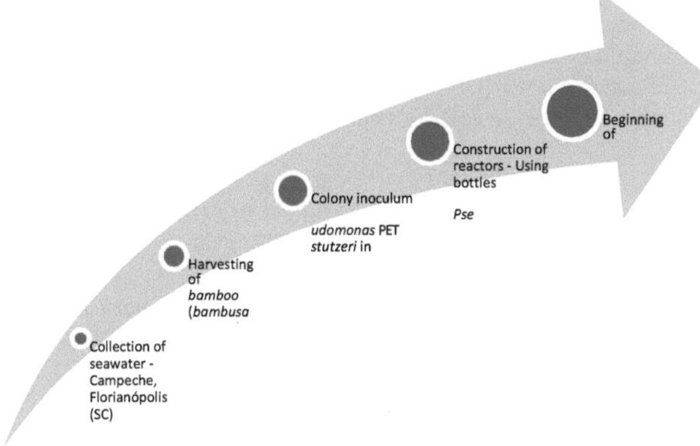

4.1 STUDY WATER

The sea water samples were collected at the coast of Campeche beach, located in the south of Florianópolis island, in Santa Catarina, as shown in Figure9.

Figure 9 - Location map of the sea water collection area.

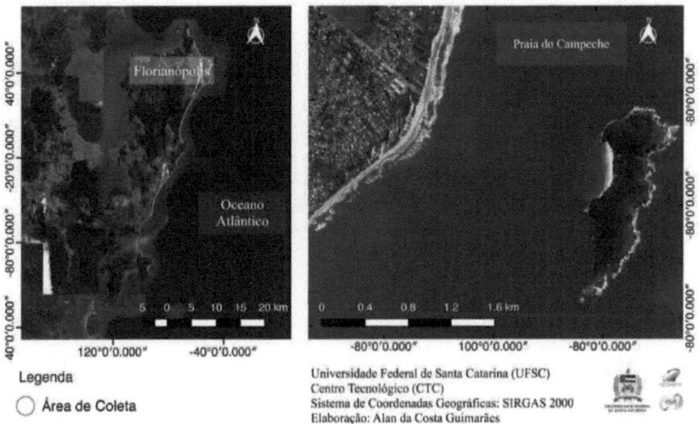

Source: Author.

The sea water samples were collected at a depth of approximately 1 meter. The samples were sent to LAPOÁ, located in the Department of Sanitary and Environmental Engineering of the Federal University of Santa Catarina (UFSC), where the reactor operation was performed and the analyses were performed.

4.2 ISOLATION, CROP IDENTIFICATION OF THE MICRO-ORGANISM *PSEUDOMONAS STUTZERI.*

The bacteria *Pseudomonas stutzeri* is found in the product Fossaklin, of the company "Bio - Brasil - Limpeza Biológica LTDA", with the composition of several microorganisms, containing:

- Bacillus subtillis 2,73x107 cfu/g;
- Pseudomonas stutzeri 6,46 x ^{107} cfu/g;
- Bacillus sp, 2,82x108 cfu/g;
- Escherichia hermanii 3,25x107 and cereal bran.

First the product was diluted with distilled water in a test tube, then, to obtain the pure culture of the bacteria of interest, the method of sowing in petri dishes was used, applying multiple streaks for isolation in solid culture medium (MADINGAN, 2016).

The culture medium was made using 8.0 grams of Soy Agar Triptone - in English: Tryptic Soy Agar (TSA), diluted in 200 mL of distilled water. The solution was poured into 10 petri dishes in a vertical laminar flow chapel after autoclaving procedures (Figures 10A and 10B), which uses saturated steam under pressure to destroy microorganisms by the combined action of temperature, pressure and humidity that promote thermocoagulation and protein denaturation (Folmer - Johnson, 1977).

Figure 10 - Materials, equipment and tests performed in microbiology laboratory.

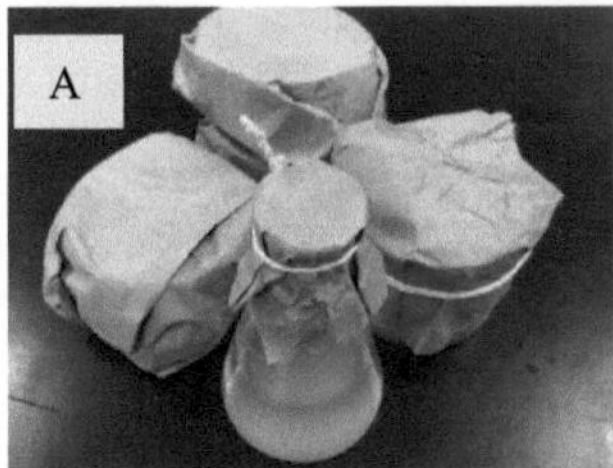

Source: Author.
Notes: A - Prepared materials for autoclave sterilization; B - Autoclave equipment used for sterilization

The striations were applied on 3 petri dishes in TSA medium and kept at room temperature for 2 days. Through the magnifying glass 5 colonies with different morphologies were selected for analysis, called colonies 1, 2, 3, 4 and 5.

Thus, to isolate *Pseudomonas stutzeri,* the application of stretch marks was carried out again from the selected colonies, until obtaining pure colonies of the microorganism in question. Therefore, the striations were applied from the selected colonies on another 5 plates in the middle of TSA, thus the morphology of the colonies was evaluated again using the magnifying glass where 8 colonies were selected for analysis called colonies 1H, 2C, 2D, 3F, 3G, 4A, 4B and 5E.

Furthermore, according to Lalucat and collaborators (2006), the strains of the
P. stutzeri can grow in starch, therefore, to favour the growth of this microorganism it has been made a culture medium with colloidal starch solution (starch gum).

1g maize starch, 0.75g Dibasic Potassium Phosphate (K2HPO4), 0.8 Magnesium Sulphate (MgSO4), 2.5g L - Glutamine and 7.5g Agar were used, diluted in 500 mL distilled water. The starch gum culture medium was poured into 25 petri dishes and 10 test tubes, and also, for the culture of the microorganism in liquid medium, culture medium was made with 12g of Soy Triptone Broth (TSB) diluted in 400 mL of distilled water, divided into 4 Erlenmeyer flasks with 100 mL of TSB in each

For the identification of the micro-organism of interest, first the Gram test was done (Figure 11A and Figure 11B), since of the micro-organisms present in the fossaklin product, *Bacillus subtillis* and *Bacillus sp.* are gram positive (KUNST, 1997). Escherichia hermani and Pseudomonas stutzeri are Gram negative (CAMPOS & TRABULSI, 2002; LALUCAT et al., 2006).

Gram staining is used to classify bacteria according to size, cell morphology and behaviour towards dyes (LEVY, 2004). After the result of the Gram test, new striations were made from the colonies identified as Gram-negative, and three solid starch gum media were used.

Figure 11 - Materials, equipment and tests performed in microbiology laboratory.

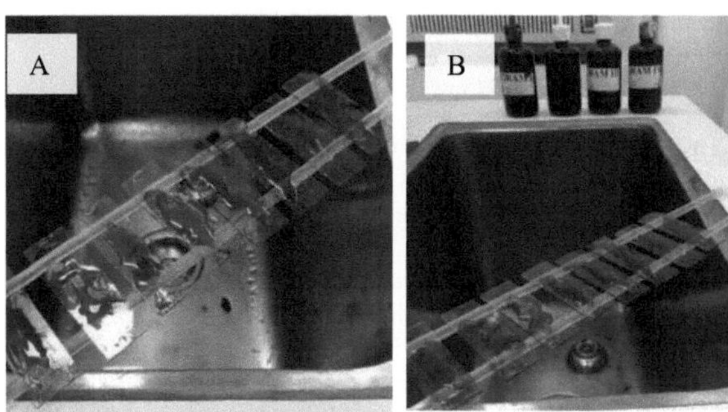

Source: Author.
Notes: A - Materials prepared for autoclave sterilization; B - Autoclave equipment used for sterilization; C and D - Performing the gram test.

The methodology used for the gram test was in accordance with the instructions of the Clinical Microbiology Manual for Infection Control in Health Services (LEVY, 2004).

The peptonated water solution was made to be added to the system as an energy source for the cultivation of microorganisms, the solution was made with 10g Peptone, 3.5g Anhydrous Monobasic Sodium Phosphate (NaH2PO4) and 1.5g Potassium Anhydrous Monobasic Phosphate (KH2PO4). All autoclave sterilization procedures were followed.

In order to identify *P. stutzeri*, tests were performed on amylase, an enzyme produced by the bacterium in question in the presence of starch. Campbell (2000), explains that starch granules normally occur as granules located in cytosol and the most common conformation of amylose is that of a helix with six residues per turn. Iodine molecules can adjust themselves within the helix to form a starch complex - iodine, which has a characteristic dark colour, the formation of this complex is a well known test for the presence of starch. Catalase, an intracellular enzyme found in *P. stutzeri* that decomposes hydrogen peroxide, has also been tested (LALUCAT *et al.*, 2006).

The pure colonies of the isolated bacteria were kept in test tubes with screw cap, kept in starch gum culture medium, kept in refrigerator at 5°C, this low temperature is used for the purpose of reducing the metabolism of microorganisms, providing the conservation of bacteria in an average period of about five to twelve months (MURRAY, 2003) . The inoculum of the bacteria isolated in four Erlenmeyers containing 100 ml of liquid culture medium TSB in each, held for 7 days before inoculum in the reactors, was performed in a laminar flow chapel.

4.3   BACTERIAL COUNT

For the bacterial count, Mc Farland's nephelometric scale was used, this scale relates the number of cells in the liquid medium from clouding or opacity in the medium, caused by the multiplication of microorganisms that oppose the free passage of light. Therefore, the greater the number of bacteria, the greater the opacity of the culture medium (LENNETE et al, 1985).

Absorbance readings, measured in a spectrophotometer, are reliable when greater than 0.05 and less than 1.2. In this way, the number of cells in a bacterial suspension can be determined roughly by comparison with McFarland scale standards.

Absorbance readings were measured before and after performing the inoculum of the bacteria. The concentration of the bacterial suspension was obtained with the aid of a spectrophotometer with a wavelength of 600 nm, thus the transmittance (A) obtained was A600 = 0.35, which according to McFarland's nephelometric scale, corresponds to approximately 1.0×10^8 colony forming units (CFU)/ml

of the culture medium (Lelliott and Stead, 1987). Thus, each ml of TSB culture medium contains 100 million cells of *Pseudomonas Stutzeri*.

4.4   BAMBOO HARVEST

The *bamboo*, of the species *Bambusa tuldoides, was* harvested in the bamboo of the Colégio Aplicação da UFSC (Figure 12A), only one bamboo stick was selected for the experiment. Oliveira (2013) explains that the kaolinitic leaves (or sheaths) give protection to the shoots on the internodes until the bamboo stalks reach full growth, and the branches begin to grow. Therefore, the presence of the kaolinitic leaves on the chosen sticks indicates that it is a stick of up to
12 months old. The young bamboo has not yet reached maturity, so it has a large amount of water, cellulose and starch in its composition (OLIVEIRA, 2013). In this way, a young bamboo stick was obtained, with the presence of kaolinitic leaves in the internodes, which means the presence of water and a high concentration of starch in its composition, favoring the cultivation of the bacteria of interest (Figure12B).

Figure 12 - *Bamboo* harvest (*Bambusa tuldoides*).

Source: Author.
Notes: A - *Bamboozal* of the species *Bambusa tuldoides*, located in Colégio Aplicação da UFSC (SC); B - Bamboo stick selected for the work, presence of kaolinitic leaves indicating that it is a stick of approximately 1 year of age.

4.5   REACTOR CONFIGURATIONS

In this study, nine types of reactors were analysed: reactors 1A and 1B, reactor 2, reactor 3, reactor 4 and reactor 5 (composed of reactors 1RB, 2 RB, 3 RB and 4 RB). The reactors operated in different conditions and configurations, according to Figure 13 and Figure 14

Figure 13 - Aerobic reactors - Reactors 1A, 1B, 2, 4 and 5 (1 RB and 2 RB).

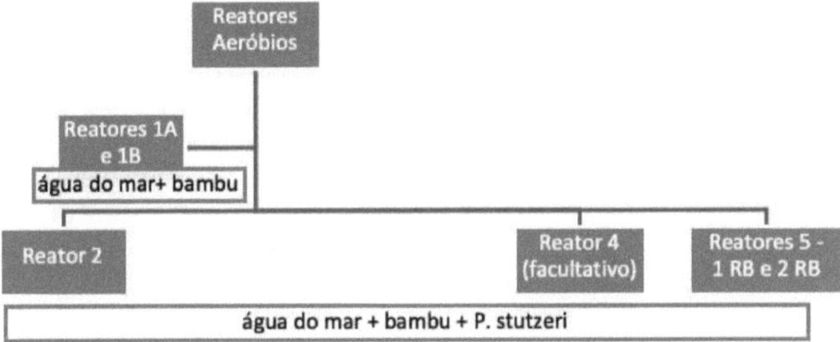

Figure 14 - Anaerobic reactors - Reactors 3, 4 and 5 (3 RB and 4 RB).

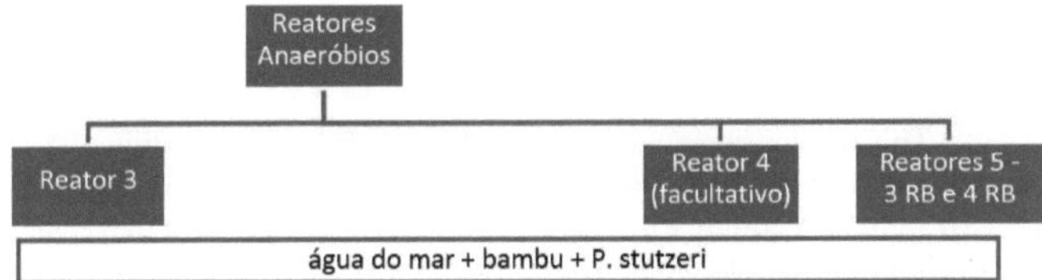

As the starch content is predominant in the inner region of the bamboo thatch wall (ALMEIDA, 2009), the knots of the bamboo pieces have been kept.

4.5.1   1A and 1B reactors

These reactors operated under aerobic conditions, are composed of 15 litres of sea water, 800g of bamboo pieces and do not contain the inoculum of *P. stutzeri*. The only differential between these reactors is the hydraulic holding time. The TDH for reactor 1A was 11 days and for reactor 1B 123 days. The dimensions of these reactors are 40 cm high by 27.6 cm in diameter. A tap was installed at the bottom of the reactor for sampling (Figure15).

Figure 15 - Illustrative scheme of Reactors 1A and 1B, composed by sea water and *bamboo* of *tuldoid bamboo* species.

Reactor 1A and 1B
Sea water + bamboo in aerobic conditions

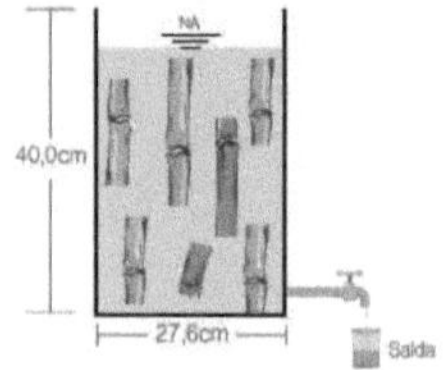

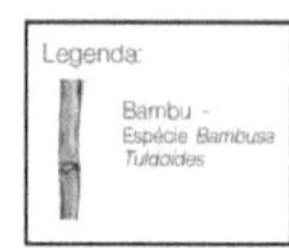

Source: Author.

4.5.2   Reactor 2

This reactor operated under aerobic conditions, consisting of 4 L seawater, 400 g *bamboo* pieces of the species *Bambusa tuldoides*, and 50 ml TSB culture medium with the inoculated *P. stutzeri* bacteria (Figure 16). Each ml of seawater contains 1.25×10^{6} CFU.

Figure 16 - Illustrative scheme of Reactor 2 composed by sea water, *bamboo tuldoid* species, and *P. stutzeri* bacteria in aerobic conditions.

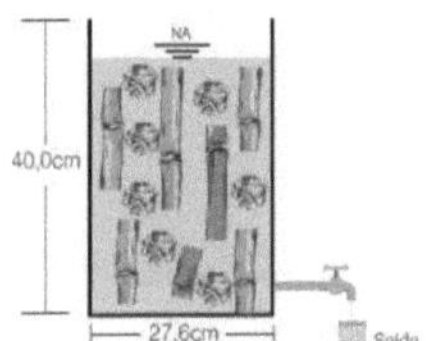

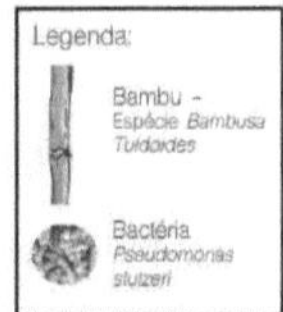

Source: Author.

4.5.3 Reactor 3

This reactor operated under anaerobic conditions, composed of 6 L sea water, 500 g of *bamboo of* the species *Bambusa tuldoides*, in the form of slats and pieces containing the nodes of the bamboo, in this reactor 100 mL of TSB culture medium containing 1.0×10^{8} CFU/ml of TSB with the bacteria *P. stutzeri* was inserted (Figure 17). Therefore, Reactor 3 contains 1.67×10^{6} CFU/ml of H2Omar.

Figure 17 - Illustrative scheme of Reactor 3, composed by sea water, *bamboo* of the species *Bambusa tuldoides*, and the bacteria *P. stutzeri* under anaerobic conditions.

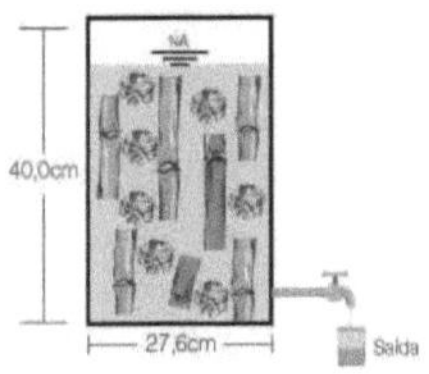

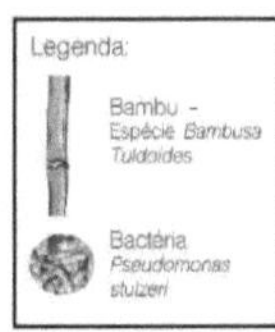

Source: Author.

The dimensions of reactors 2 and 3 are: 40 cm high by 27.6 cm in diameter. A tap has been installed at the bottom of the reactor to collect the samples. The electrical conductivity and pH monitoring of reactors 2 and 3 were performed for a total period of 55 days.

For reactors 2 and 3, the analysis of the apparent colour and turbidity parameters was also carried out. Also, the chloride determination analysis was performed for reactor 3 sample.

4.5.4   Reactor 4

This reactor operated in batch flow in series, consisting of 3 chambers, the first chamber operated in aerobic condition, the second in optional condition and the third chamber in anaerobic condition.

Reactors 2 and 3 acted as *start - up reactors* for the chambers of reactor 4. 700 ml of raw seawater and 300 ml of each of the start - up reactors with 7-day TDH for reactor 2 (aerobic) and 11 days for reactor 3 (anaerobic) were thus inserted.

The water from reactor 2 was inoculated in chamber 1 and the water from reactor 3 was inserted in chambers 2 and 3, as shown (Figure 18).

Figure 18 - Illustrative diagram of Reactor 4, operating in batches and series, containing three chambers with sea water, bamboo and *P.stutzeri* bacteria: chamber 1 (aerobic condition), chamber 2 (optional condition), chamber 3 (anaerobic condition).

Reactor 4
Sea water + bamboo + *P. stutzeri*
in serial batch

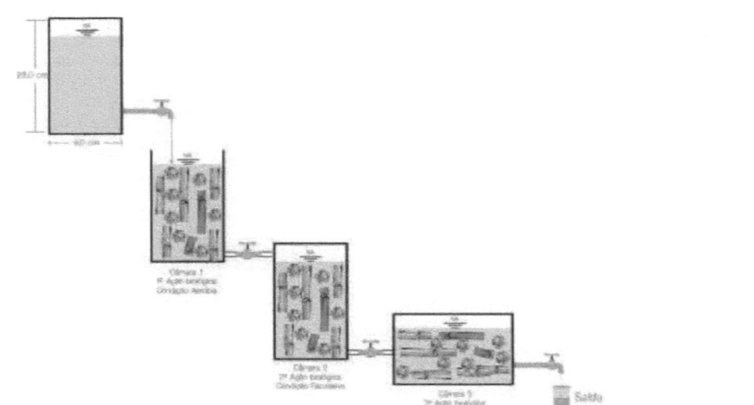

Source: Author.

The reactor chambers 4 are detailed below:

- Camera 1 - This is the first stage of the biological action of the system, this camera consists of a PET bottle of 1.5 l, 282.51 g of bamboo, one piece with knot and another divided into four slats, TDH of 13 minutes, in aerobic conditions. In this chamber, 300 ml of seawater from Reactor 2 has been inoculated. As Reactor 2 has a concentration of 1.25×10^6 CFU/ml of H2Omar, 3.75×10^8 CFU are contained in these 300 ml of seawater. When diluted in 700 ml of raw seawater, the concentration of *P. stutzeri* UFC is 5.36×10^5 UFC/ml of H2Omar.

- Camera 2 - The second stage of the biological action of the system occurs in camera 2 in an optional condition, this camera contains colonies of the bacteria *Pseudomonas stutzeri*, 261.40 g of bamboo, one piece with knot and another divided into four slats and a TDH of 5 minutes. In this chamber, 300 ml of seawater from Reactor 3 has been inoculated, as Reactor 3 has a concentration of 1.67×10^6 CFU/ml of H2Omar, 5.01×10^8 CFU are contained in these 300 ml of seawater. When diluted in 700 ml of raw seawater, the concentration of *P. stutzeri* UFC is 7.16×10^5 UFC/ml of H2Omar.

- Camera 3 - This represents the last stage of the biological action of the system, it is a 1.5 l PET bottle chamber with 232.26 g of bamboo, one piece with knot and another divided into eight slats with 7-minute TDH in anaerobic conditions. In this chamber, 300 ml of seawater from reactor 3 was inoculated, thus containing the concentration of 7.16×10^5 CFU/ml of H2Omar, as well as chamber 2.

6 mm holes were drilled in the sides of the bamboo in order to facilitate the passage of water inside (

46

Figure 19C). Figure 19 shows the materials for the beginning of operations.

Figure 19 - Materials for starting operations.

Notes: A - Reactor 1 (aerobic start - up reactor); B - Reactor 2 (anaerobic start - up reactor); C - Pieces of bamboo inserted in the anaerobic start - up reactor, in the form of slats and pieces keeping the bamboo nodes with holes in the sides; D - *Pseudomonas stutzeri* inoculated in TSB culture medium.

4.5.5   Reactors 5

These are the batch reactors (RB), consisting of four isolated reactors where 20 ml of TSB culture medium containing 1.0×10^{8} CFU/ml of *Pseudomonas stutzeri* suspended cells in 250 ml of seawater has been added, and as an energy source for the micro-organisms, 20 ml of peptonated water has been added to each of the four reactors. Thus each reactor contains 8.00 x ^{106} CFU/ml of H2Omar.

What differs these reactors are the oxygen conditions and the bamboo support medium, thus the reactors:

- Reactors 1 RB and 2 RB contain sea water and the bacteria inoculated under aerobic conditions, and reactor 1 RB has 39.3g of the bamboo support medium (pieces in discs);

- Reactors 3 RB and 4 RB have seawater and the bacteria inoculated under anaerobic conditions, the bamboo support medium (50.2g, containing the bamboo node) was inserted only in reactor 3 RB.

In order to evaluate the variation in electrical conductivity before and after the filtration process, successive filtrations were performed on all reactors in this set, using membranes with porosity of 0.45 μm and 0.22 μm, as shown in Figure20.

Figure 20 - Illustrative diagram of reactor 5, closed system reactors, varying aerobic and anaerobic conditions and the bamboo support medium.

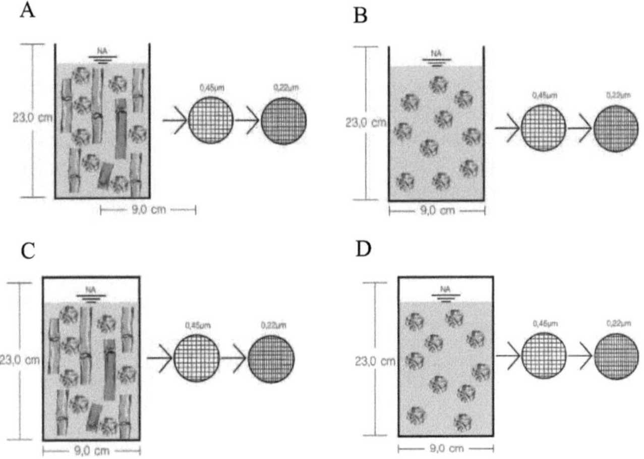

The relationship between the reactor type and the CFU/ml concentration of H2Omar is shown in Table 6.

Table 6: CFU/ml H2Omar concentration by reactor type.

Reactor Type		CFU/ml concentration of H2Omar
Reactor 2		$1{,}25 \times 10^6$
Reactor 3		$1{,}67 \times 10^6$
Reactor 4	Camera 1	$5{,}36 \times 10^5$
	Camera 2	$7{,}16 \times 10^5$
	Chamber 3	$7{,}16 \times 10^5$
Reactor 5		$8{,}00 \times 10^6$

Source: Author.

4.6 ANALYTICAL PROCEDURE

All the methodologies used to determine the physical and chemical parameters of seawater were based on the procedures described in the *Standard Methods for Examination of Water & Wastewater (APHA*, 2012) and the practical manual for water analysis (FUNASA, 2006). The physical-chemical parameters that have been used for the characterisation of raw and treated study water can be observed in the
Table 2.

Table 2 - Physical - chemical parameters used in the characterization of raw and treated study water.

Parameter/ Unit	Method	Equipment
Apparent colour (uH)	2120C (APHA, 2012)	DR Spectrophotometer 2800 HACH
pH	4500 H+ (APHA, 2012)	pHmetroHACH Model 8306

Turbidity (uT)	2130 B (APHA, 2012)	TurbidimeterHach Model 2100p
Conductivity (mS cm-1)	2510 B (APHA, 2012)	AZ Pipeline - Model HQ10d

For the filtration step, cellulose nitrate filter membranes of 0.45 μm and 0.22 μm and a vacuum filtration system of 500 mL were used.

For the determination of chlorides, the method used was silver nitrate titration, following the methodology indicated in the practical manual for water analysis (FUNASA, 2006). A standard solution of Silver Nitrate 0.014 N and an indicator solution of Potassium Chromate (K2CrO4) were used. With the pH of the samples between 7 and 9, 1 mL K2CrO4 was added to each sample.

The Silver Nitrate 0.014 N Standard solution has been titrated to colour change to reddish yellow, end point of the titration. Filtration was carried out on membranes with porosity of 0.45 μm and 0.22 μm before starting the titration so that colour and turbidity would not interfere with the results of the analyses. A blank was made in the same way as the samples. The calculation for the determination of chlorides is presented in Equation 4.

$$\underset{103}{Cl}\ (mg/L) = \frac{(\text{Æ-B) s N s 35,45 s}}{\text{NL of ANOCTRA}}$$

(4)

Where:
A = mL of the titrant spent on the sample; B = mL of the titrant spent on the blank;
N = Normality of the holder

The chloride determination analysis was only performed on a sample of raw seawater and reactor 3, as it was the reactor that showed the best result in terms of reduction of electrical conductivity. The sample of reactor 3 was diluted 200 times, that is, 0.5 mL of sample was used for 99.5 mL of distilled water. This analysis was done in duplicate.

The diagram in Figure 21 illustrates in a simplified way the parameters analysed in the respective reactors.

Figure 21 - Parameters analysed in the respective reactors.

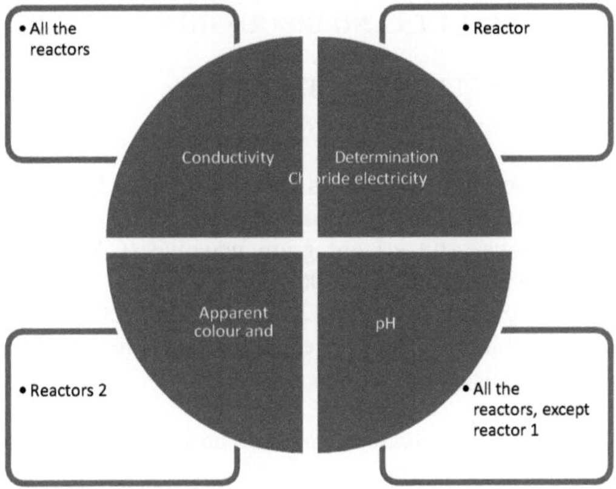

5 RESULTS AND DISCUSSION:

5.1 SECLUSION AND CULTURE PSEUDOMONAS STUTZERI

Of the micro-organisms present in fossaklin, Bacillus subtillis and Bacillus sp. are gram positive (KUNST, 1997). Escherichia hermani and Pseudomonas stutzeri are gram negative (CAMPOS & TRABULSI, 2002; LALUCAT et al., 2006). Thus, it was
First the gram test (Figure 22A and Figure22D) was performed on the selected colonies which revealed that only the 3G, 4A and 4B colonies were gram negative.

Figure 22 - *Pseudomonas stutzeri* isolation and identification results.

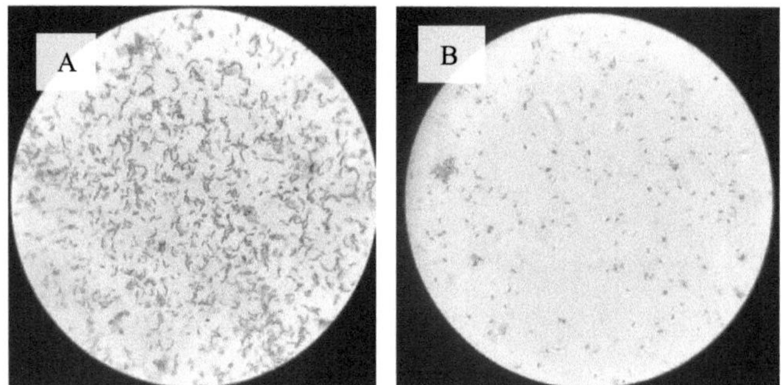

Source: Author.
Notes: A and B - 4A and 4B plates gram negative cologne, respectively.

According to Oliveira and collaborators (2006), the activity of amylase can be detected from the addition of iodine solution, which is able to stain the starch blue, and the regions where the polysaccharide (starch) has been degraded by amylase do not occur to the heart, which shows the activities of the enzyme in the form of a colorless halo.

Thus, after striations were made on the 3G, 4A and 4B plates in starch gum culture medium, these were kept at room temperature for 1 day, then, to detect the enzyme activity, iodine solution was added to the culture medium, thus revealing the presence of lighter regions around the microorganisms, these regions correspond to the places where the enzyme degrade the starch (Figure 23A) all the plates analysed reacted with the iodine.

Lalucat et al. (2006) cites that *P. stutzeri* presents positive catalase tests, therefore, the microorganism has an enzyme that breaks down hydrogen peroxide molecules (H2O2) into water and oxygen, thus, the dose of two drops of hydrogen peroxide in each plate (3G, 4A and 4B) was used, where the formation of bubbles occurred immediately, conferring the positive result to catalase (Figure23B).

Figure 23 - *Pseudomonas stutzeri* isolation and identification results.

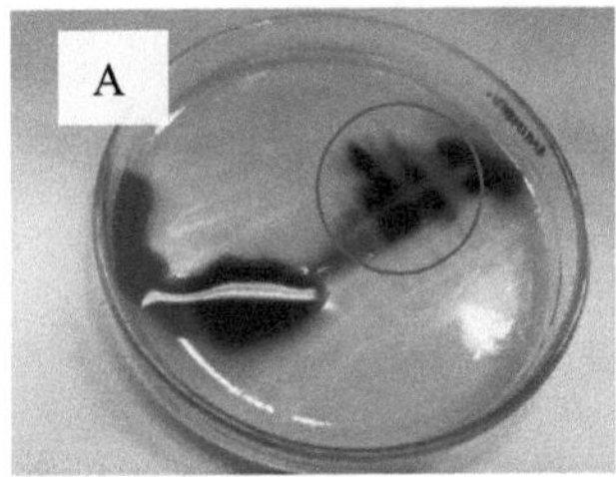
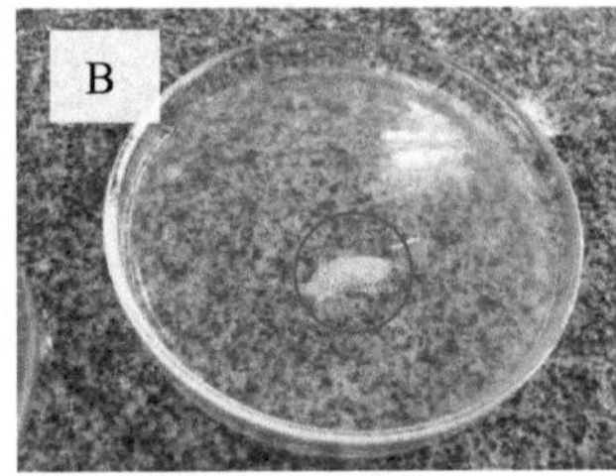

Notes: A - Amylase production; B - Catalase production revelation.

5.2   REACTOR EFFICIENCY

5.2.1   1A and 1B reactors

According to Figure 24, it can be seen that in Reactor 1A, with an 11-day TDH, the electrical conductivity increased by 3.97%, varying from 47.9 mS to 49.8 mS (Figure 24).

Figure 24 - Variation of electrical conductivity of reactor 1A, composed of sea water and *bamboo* of the species *Bambusa tuldoides*, in aerobic conditions with TDH of 11 days.

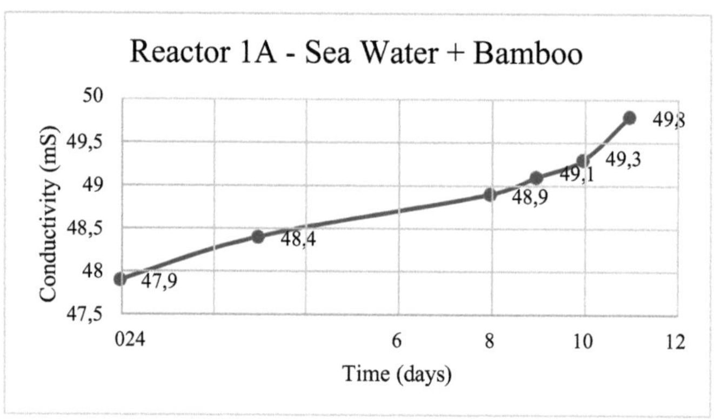

In the 1B Reactor, with a TDH of 123 days, the increase in electrical conductivity from 47.9 to 70 mS (increase of 46.14%) was again verified (Figure 25).

56

Figure 25 - Electrical conductivity variation of the 1B reactor, composed of sea water and *tuldoid bamboo*, in aerobic conditions with 123 day TDH.

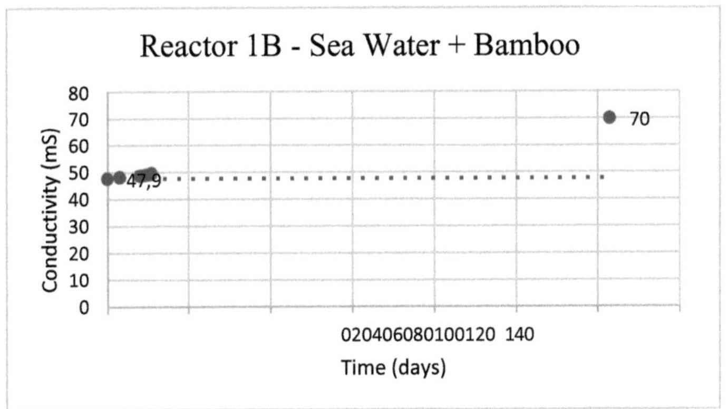

As Reactor 1 operated in a closed system, the substances in the bamboo may have been carried into the study water, so they remained in the system. According to Muller (2016), the increase in electrical conductivity may indicate the passage of substances from bamboo into the produced water.

In addition, the degradation of organic matter generates the release of H+ protons which can be bound with other elements present in seawater and form some acids such as HCL, since the chlorine atom reaches the octet structure by sharing one of its valence electrons with the hydrogen atom (VOLLHARDT et al., 2013).

As for biofilm generation, the 1A and 1B reactors were the ones that generated the thickest biofilms, Figure 26

Figure 26 shows the bifilm cycle in these reactors, showing their four generation stages: surface fixation with 4-day DTH (Figure 26A), adhesion and cell proliferation with 8-day DTH (Figure 26B), maturation with 22-day DTH (Figure 26C) and deployment with 35-day DTH (Figure 26D) (PROAL, 2008). No thick biofilms were recorded in anaerobic reactors.

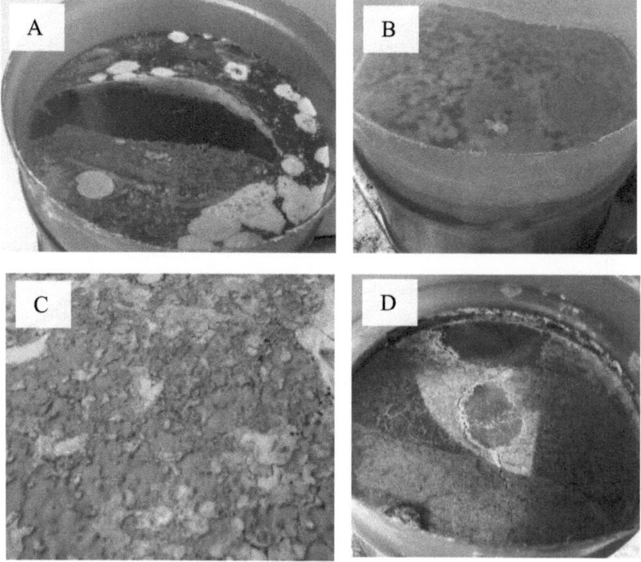

5.2.2   Reactors 2 and 3:

According to Figure 27 and Figure28, it can be seen that in Reactor 2, with TDH of 55 days, the electrical conductivity increased from 52.7 mS to 54.4 mS (3.23 %), and the pH from 7.99 to 8.41 (5.26 %).

Reactor 3, with a TDH of 62 days, the electrical conductivity was reduced from 52.7 mS to 49.9 mS (5.31 %) and the pH decreased from 7.99 to 6.85 (14.27 %)

After 26 days the electrical conductivity started to increase in the aerobic system (Reactor 2), while in the anaerobic system (Reactor 3) the electrical conductivity was subtly reduced until day 20, then it kept
- The number of days of operation was constant and showed little variation until the end of the 55-day operation (Figure 27).

58

Figure 27 - Variation of electrical conductivity of reactors 2 and 3, which consist of seawater, pieces of bamboo and *P. stutzeri* bacteria, under aerobic and anaerobic conditions, respectively.

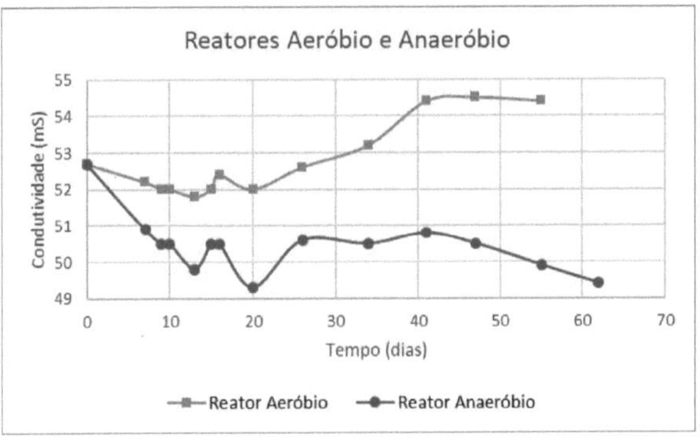

Figure 28 - Electrical conductivity and pH variation of reactors 2 and 3, which consist of seawater, pieces of bamboo and *P. stutzeri* bacteria, under aerobic and anaerobic conditions, respectively.

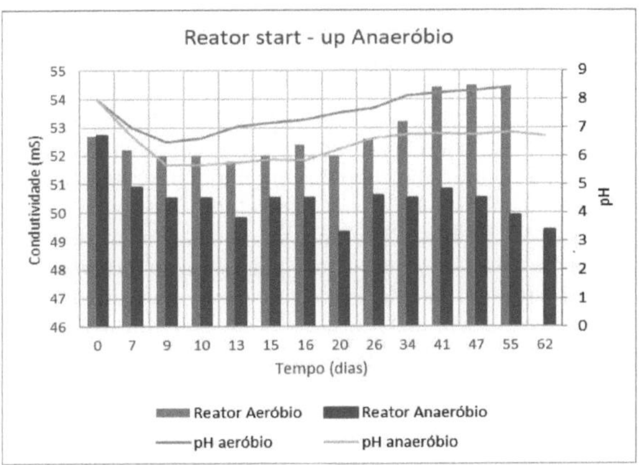

Pseudomonas stutzeri is positive for amylase, i.e. it produces extracellular enzyme that degrades complex organic matter (starch) in simple assimilable molecules (glucose). According to Facincani (2002), this bacterium follows the metabolic pathway of Entner - Doudorof, described only in microorganisms, being the main route of glucose degradation in several Gram-negative bacteria, including *Pseudomonas*. The net result of glucose degradation by this route is the formation of two pyruvate molecules (C3H4O3) or pyruvic acid, composed of three carbons (FACINCANI, 2002).

According to Campbell (2000), the pyruvate has many destinations when formed, and when the metabolism is aerobic, the pyruvate loses carbon dioxide, and the two remaining carbon atoms are bound to coenzyme A as an acetyl group, forming acetyl-CoA, which then enters the citric acid cycle (CAMPBELL, 2000).

As the amount of pyruvate is limited in the cell, it needs to be regenerated and from the anaerobic glycolysis, the pyruvate can be reduced in lactic acid (lactate) (FACINCANI, 2002). As the acids form ionic solutions, i.e. they generate free ions, the conductivity of electrical current is guaranteed.

When in anaerobiosis, *Pseudomonas stutzeri* does not ferment, it therefore presents only anaerobic respiration and the complete Krebs cycle, NADH being the initial electron donor of the anaerobic electron transport chain (FACINCANI, 2002). During the process of anaerobic respiration of *Pseudomonas stutzeri* various substrates are generated, which with contact with seawater different bonds can be formed.

As for the determination of chlorides, the calculations performed were determined from Equation 4 of item 4.6. Where N = 0.014 and B = 1.3 mL. Raw sea water had a chloride concentration of 19.99 (g/L Cl), where A = 21.3 mL. The chloride concentration in Reactor 3 was 15.24 (mg/L Cl) with A = 16.55 mL. Therefore, Reactor 3 had a chloride removal of 23.75%, which means the removal of dissolved salts, and explains the reduction of electrical conductivity in this reactor.

In relation to the apparent colour and turbidity parameters measured in reactors 2 and 3, in accordance with Ministry of Health Order No. 5/2017, the VMP for apparent colour is 15 uH.

Intake water was 23 uH, reactor 2 (aerobic) 1643 uH and reactor 3 (anaerobic) 552 uH. Therefore, it is concluded that none of the samples presented values within the maximum permitted values of potability established for apparent colour (Figure 29).

Figure 29 - Variation of the apparent colour (uH) of the incoming raw water and reactors 2 and 3, after 55 days of operation.

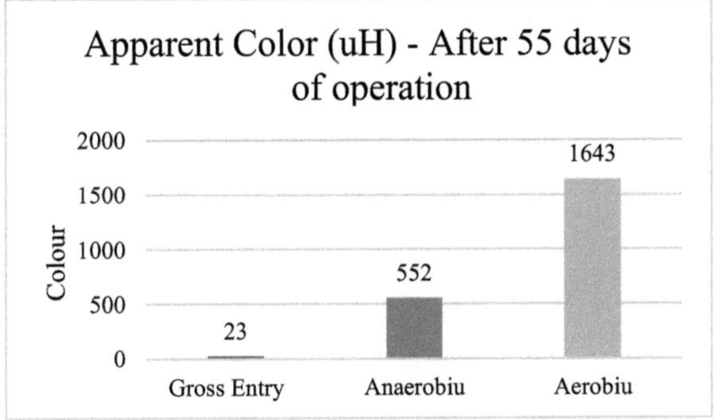

Regarding turbidity, the samples of reactors 2 and 3 are above the maximum allowable values (5 uT) established by Ministry of Health Ordinance No. 5/2017, with 66.5 uT and 19.2 uT, respectively (Figure 30). Therefore, only the raw water input (1.05 uT) complies with the organoleptic standards of potability.

Figure 30 - Variation of turbidity (uT) of raw input water and reactors 2 and 3, after 55 days of operation.

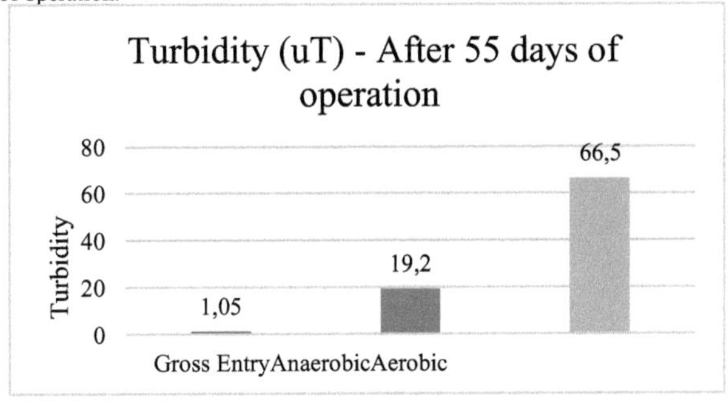

Figure 31 shows the visual comparison of the apparent colour and turbidity of the samples of reactors 2 and 3 and of the incoming raw water.

Figure 31 - Comparison of colour and turbidity of samples after 55 days.

Source: *Author.*

As for the increase in apparent colour and turbidity recorded in reactors 2 and 3, Limas (2009) explains that bamboo expels a resin in the first days of use, increasing the amount of colour in the water and that this process can take 90 days to cease. In addition, all the reactors showed a strong odour, being even more intense for the samples of anaerobic reactors containing the bamboo.

5.2.3   Reactor 4

Reactor 4 operated with a 120 minute TDH, from Figure 32 it is possible to observe that there was a variation in electrical conductivity from 52.3 mS to 50.2 mS, representing a reduction of 4.01%, the pH ranged from 7.95 to 6.14 (reduction of 22.76%) (Figure 32).

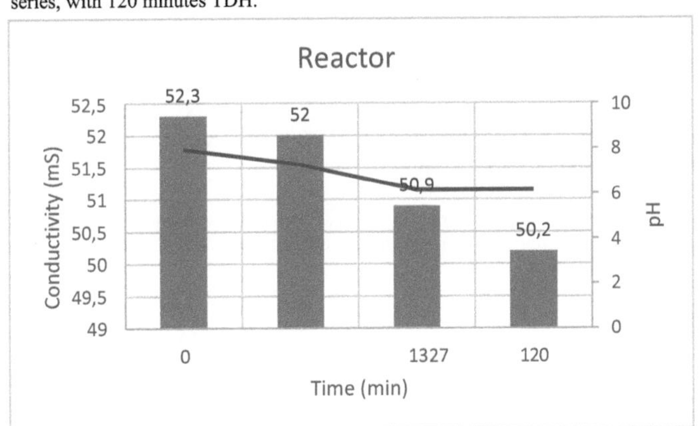

Reactor 4 was the lowest pH of all the reactors analysed, acid formation can be the reason for the reduction of pH in both aerobic and anaerobic reactors.

In general, all reactors comply with the pH tolerance conditions for *P. stutzeri* cultivation, since acid conditions are not tolerated by these microorganisms and growth is inhibited with a pH of 4.5 (LALUCAT et al., 2006). Furthermore, the presence of CO_2 and highly alkaline ions such as sodium, potassium and calcium ions tend to make seawater slightly alkaline with values between 7.5 and 8.4 (RESENDE, 2009).

5.2.4   Reactor 5

In reactor 1 RB, electrical conductivity was reduced by 2.13% (ranging from 51.6 mS to 50.5 mS), after filtration the electrical conductivity was 49.9 mS (reduction of 3.29%) (Table 7).

The reactor 2 RB, had a reduction in electrical conductivity from 51.6 mS to 51.2 mS (variation of 0.77%). After the filtration step, the electrical conductivity decreased to 50.2 mS (reduction of 2.71%), which represents a 6-day reduction of 1.95% in the electrical conductivity of the sample with DTH after filtration (Table7).

The 3 RB reactor had its electrical conductivity reduced from 51.6 mS to 49.5 mS (reduction of 4.07%), after filtration the electrical conductivity was reduced to 49.0 mS representing a total reduction of 5.04% (Table7).

In reactor 4 RB, the electrical conductivity was reduced from 51.6 mS to 49.7 mS (representing a reduction of 3.68%), after filtration the electrical conductivity decreased to 48.9 mS, reactor 4 RB showed a total reduction of 5.23% (Table7).

Table 7 - Percentages of electrical conductivity reduction efficiencies in relation to the output sample (6-day DTH), and after filtrations, as well as total system efficiency (after 6-day DTH and filtrations).

Configuration	Efficiencies - electrical conductivity (%)		
	Removal efficiency (%) after 6 days	Removal efficiency (%) after 6 days and filtration (0.45 and 0,22 μm)	E (%) total
1 RB - Aerobic Bamboo-free	2,13	1,19	3,29
2 RB - Aerobic With Bamboo	0,77	1,95	2,71
3 RB - Anaerobic Without Bamboo	4,07	1,01	5,04
4 RB - Anaerobic With Bamboo	3,68	1,61	5,23

In relation to pH, the 3 RB reactor had the lowest pH (6.35), as shown in Figure 33.

Figure 33 - pH variation of reactor 5, batch reactor with 6-day TDH, presenting the result of the entry and exit of the systems under aerobic and anaerobic conditions, containing or not the bamboo support medium.

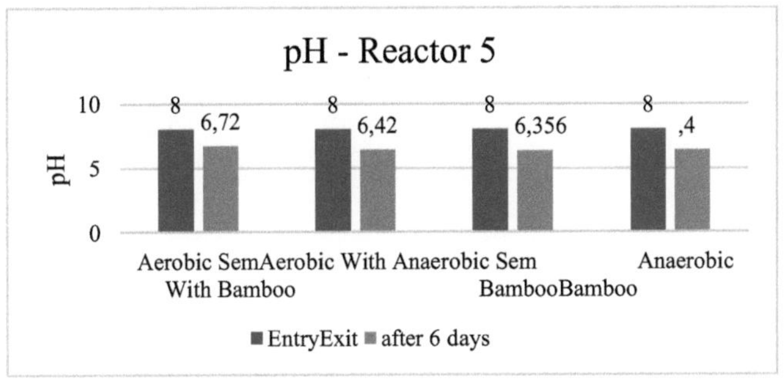

Among the reactor 5 set, reactor 4 RB showed the greatest reduction in electrical conductivity (5.29%) composed of bamboo support medium, 6-day TDH and successive membrane filtrations with porosity of 0.45 μm and 0.22 μm (Table 7). This reactor had its pH reduced from 7.95 at inlet to 6.40 after 6 days of hydraulic holding, indicating that possible acid formations occurred in this reactor.

Thus, in reactor 5 the possibility of dissolved salts being absorbed or adsorbed by the bacterium in question was studied, where successive filtrations were made on membranes with porosity of 0.45 μm and 0.22 μm. The absorption of dissolved salts is hardly the factor in the reduction of electrical conductivity since, following the reasoning of the pump mechanism - potassium in the cells, the potassium ions are moved into the cell as simultaneously the sodium ions are moved out of the cell. In this way, the resting potential of the cell is maintained (negative in the intracellular medium and positive in the extracellular medium) (CAMPBELL, 2000). Furthermore, the sodium concentration is very low in the cell composition of a bacterium, which is only 0.5% to 1.0 % (MADIGAN, 2016). Much higher concentrations of this bacteria could cause a cellular imbalance.

Figure 34 shows the variation of electrical conductivity in reactor 5 as a whole, both after the hydraulic holding time of 6 days and after filtration.

Figure 34 - Variation of electrical conductivity of reactor 5, a 6-day TDH batch reactor, presenting the result of the input and output of the reactors under aerobic or anaerobic conditions, containing or not the bamboo support medium, as well as the electrical conductivity after the filtration steps in filter membranes with porosity of 0.45μm and 0.22μm, respectively.

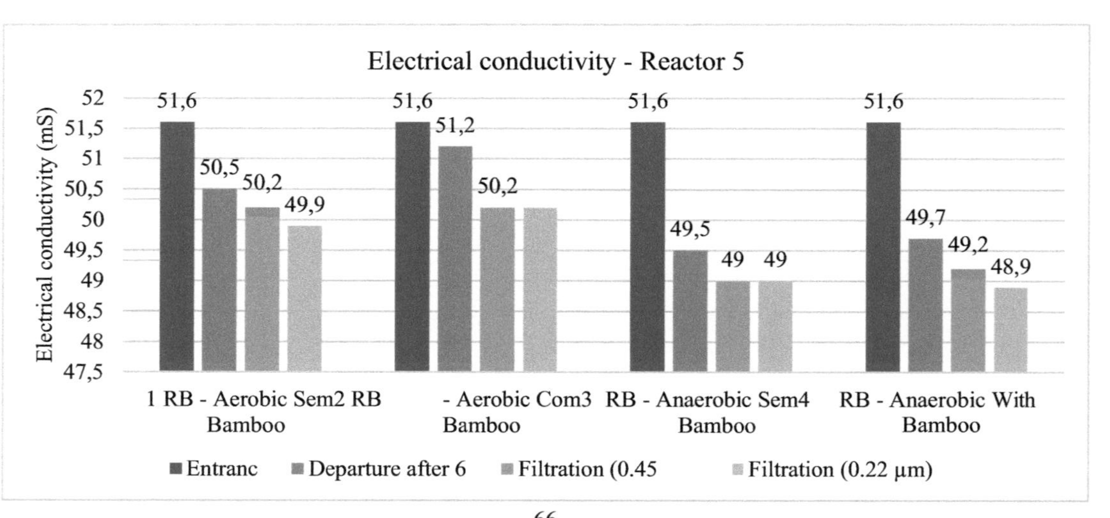

Pilling (2015) explains that although there are micro-organisms that survive in high concentrations of dissolved salts, these micro-organisms do not consume salt themselves, but have developed survival mechanisms in these types of environments, called osmoadaptation. Therefore, in order to understand the removal of chlorides by means of bacterial metabolism, future studies should be prepared to contribute to the understanding of this process.

Although it is possible to interpret the behaviour of *Pseudomonas stutzeri* as regards the degradation of starch in aerobic and anaerobic environments, several other microorganisms are found in seawater as well as in bamboo. Therefore, different types of substrates are generated in the degradation processes of organic matter present in the reactors, making the analysis of the systems complex, since the substrates formed are not only generated from the metabolism of *Pseudomonas stutzeri*.

Table 8 summarizes the operating conditions and the results of chloride determination, electrical conductivity and pH of all reactors analyzed.

Table 8 - Results of pH and electrical conductivity and chlorides in all analyzed systems.

		Condition	TDH	CFU concentration of *P.stutzeri/ml* of H2Omar	Contains Bamboo	Removal of Chlorides (%)	pH Entrance	pH Exit	Electrical Conductivity Variation (%)
Reactor 1A		Aerobics	123 days	-	Yes	-	-	-	+46,14
Reactor 1B		Aerobics	11 days	-	Yes	-	-	-	+3.97
Reactor 2		Aerobics	55 days	$1,25 \times 10^{6}$	Yes	-	7,95	8,41	+3,23
Reactor 3		Anaerobia	62 days	$1,67 \times 10^{6}$	Yes	23,75	7,99	6,7	-6,26
Reactor 4		Aerobic / Anaerobic	120 min	$^{1} - 5,36 \times 10^{5}$	Yes	-	7,95	6,14	-4,02
				$^{2} - 7,16 \times 10^{5}$	Yes	-			
				$^{3} - 7,16 \times 10^{5}$	Yes	-			
Reactor 5	1 RB	Aerobics	6 days	$8,00 \times 10^{6}$	Yes	-	7,95	6,72	-3,29
	2 RB	Aerobics	6 days	$8,00 \times 10^{6}$	No	-	7,95	6,42	-2,71
	3 RB	Anaerobia	6 days	$8,00 \times 10^{6}$	No	-	7,95	6,35	-5,04
	4 RB	Anaerobia	6 days	$8,00 \times 10^{6}$	Yes	-	7,95	6,40	-5,23

NOTE - 1 for camera 1, 2 - for camera 2 and 3 for camera.

6 CONCLUSION

It is concluded that all the reactors analysed did not have satisfactory results with respect to apparent colour, turbidity. However, the reduction of electrical conductivity and chloride concentration in anaerobic reactors was evidenced.

In aerobic reactors, the parameters apparent colour, turbidity and reduction of electrical conductivity have the results negatively aggravated in relation to the input samples. Electrical conductivity analyses show that this parameter starts to increase from the 26th day of operation in aerobic reactors, and in Reactor 1, which *Pseudomonas stutzeri* was not inoculated, the electrical conductivity already increases from the first day of operation.

In 6-day TDH, both aerobic and anerobic reactors containing the bamboo showed reduced pH and electrical conductivity. The use of bamboo support medium resulted in a further reduction of electrical conductivity under anaerobic conditions. Anaerobic reactors compared to aerobic reactors showed better results regarding colour, turbidity and reduction of electrical conductivity. Anaerobic and anaerobic reactors presented better electrical conductivity reduction efficiencies for a TDH of up to 20 days.

In anaerobic reactors, a subtle reduction in electrical conductivity, with a maximum reduction of 6.26%, was recorded. This result may be due, in parts, to a standard deviation. Despite this, the chloride determination analysis carried out in reactor 3 shows that the chloride concentration was reduced by almost 24%. This result reveals the removal of dissolved salts and may explain the reduction in electrical conductivity in this reactor.

The analysis of the performance of *Pseudomonas stutzeri* in constructed systems is of complex analysis, since several substrates are formed from the degradation of organic matter contained in the reactors, a consequence of the metabolic processes of the numerous microorganisms in seawater and bamboo.

Research aimed at reducing chlorides by microorganisms is recommended for a better understanding of how this parameter was reduced in the reactor analysed.

7   REFERENCES

ANA - National Water Agency. Water, facts and trend. ANA, (2009). **Brazilian Business Council for Sustainable Development**. Available at   in: <http://arquivos.ana.gov.br/imprensa/publicacoes/fatosetendencias/edica o_2.pdf> Accessed: July 2019.

APHA, Awwa. Wpcf. (2012) Standard methods for the examination of water and wastewater. **American Public Health Association, Washington**.

AZZINI, A. et al. Variations in the cellulose fibre and starch contents of bamboo stalk. **Bragantia, Campinas-SP**, v. 46, n. 1, p. 141-145, 1987.

BELILA, A. et al. Bacterial community structure and variation in a full-scale seawater desalination plant for drinking water production. **Water research**, v. 94, p. 62-72, 2016.

BOSCH R.; ROS, REDIGOR J.L.; ESTELLÉS, R.M. Biocleaning of nitrate alterations on wall paintings by Pseudomonas stutzeri. **International Biodeterioration & Biodegradation, v.** 84, p. 266-274, 2013.

BUSCH, M.; MICKOLS, W. E. Reducing energy consumption in seawater desalination. **Desalination**, v. 165, p. 299-312, 2004.

CAMPBELL, MARY K. Biochemistry. 3rd edition. **Porto Alegre: Artmed Editora**, p. 418 - 420, 2000.

CAMPOS, L.C.; TRABULSI, L.R. Escherichia. In.: TRABULSI, L.R. et al. **Microbiology**. 3 ed. São Paulo : Atheneu, 2002, p.215-228.

ACADEMIC TECHNOLOGY CENTRE OF THE INSTITUTE OF UFRGS PHYSICS (CTA - IF/UFRGS). Eye On Water: Conductivity Meter for Environmental Monitoring Purposes. Available at: http://cta.if.ufrgs.br/projects/medidor-de-condutividade-eletrica-environmental monitoring/wiki/Estado_da_Arte. Apr. 2019.

CONAMA, Resolution. 357, of March 17, 2005. **National Environment Council-CONAMA**, v. 357, 2005.

ELIMELECH, M; PHILLIP, W. A. The future of seawater desalination: energy, technology, and the environment. **science**, v. 333, n. 6043, p. 712-717, 2011.

FACINCANI, A. P. Estudo de isoenzimas de metabolismo de carboidratos e clonagem e expressão da enolase de Xylella fastidiosa. 2002.

FOLMER - JOHNSON TNO. Introduction to thermology. São Paulo: Nobel; 1977. p. 67.

FORRESTAL, C. et al. Microbial desalination cell with capacitive adsorption for ion migration control. **Bioresource technology**, v. 120, p. 332-336, 2012.

FRITZMANN, C. et al. State-of-the-art of reverse osmosis desalination. **Desalination**, v. 216, n. 1-3, p. 1-76, 2007.

FUNASA, FN d S. **Practical manual for water analysis**. 2006.

GE, Z. et al. Effects of draw solutions and membrane conditions on electricity generation and water flux in osmotic microbial   fuel cells. **Bioresource technology**, v. 109, p. 70-76, 2012.

GUDE, V.; KOKABIAN, B.; GADHAMSHETTY, V. Beneficiary bioelectrochemical systems for energy, water, and biomass production. **Journal of Microbial & Biochemical Technology, v.** 6, p. 2, 2013.

HARRISON, J. J. et al. Biofilms: a new understanding of these microbial communities is driving a revolution that may transform the science of microbiology. **American Scientist**, v. 93, n. 6, p. 508-515, 2005.

INGALLS, A. E. et al. Quantifying archaeal community autotrophy in the mesopelagic ocean using natural radiocarbon. **Proceedings of the National Academy of Sciences**, v. 103, n. 17, p. 6442-6447, 2006.

Natural effluent treatment system. **BR n. PI0704292-2**, 23 Nov. 2007, 21 July 2009.

JACOVIDE, C.; PARVIZI, J. Biofilm infections: Newer understanding of old problem. **Orthopedics Today September**, 2010.

KIM, Y.; LOGAN, B. E. Microbial desalination cells for energy production and desalination. **Desalination**, v. 308, p. 122-130, 2013.

KIRCHMAN, D. L.; GASOL, J. M. (Ed.). **Microbial ecology of the oceans**. New York: Wiley-Liss, 2000.

KUNST, F. et al. The complete genome sequence of the gram-positive bacterium Bacillus subtilis. **Nature**, v. 390, n. 6657, p. 249, 1997.

LALUCAT, J. et al. Biology of Pseudomonas stutzeri. **Microbiol. Mol. Biol. Rev.** v. 70, n. 2, p. 510-547, 2006.

LEE, J.D. Inorganic chemistry not so concise. Translation of the 5th English edition: Henrique E. Toma, Koiti Araki, Reginaldo C. Rocha - São Paulo: **Edgard Blucher**, 1999.

LELLIOTT, R. A. et al. Methods for the diagnosis of bacterial diseases of plants. **Blackwell Scientific Publications**, 1987.

LEVY, C. E. et al. Manual of clinical microbiology for infection control in health services. **Brasília: Editora Agência Nacional de Vigilância Sanitária**, 2004.

LENNETE, E. H.; HANSLER, J. R.; SHADOMY, H. J. **Manual of Clinical Microbiology**. Washington. 1985.

LIBES, S.M. Na **Introduction to Marine Biogeochemistry**. New York, John Wiley & Sons. 734 p., 1992.

LÓPEZ, D.; VLAMAKIS, H.; KOLTER, R. Biofilms. **Cold Spring Harbor perspectives in biology**, v. 2, n. 7, p. a000398, 2010.

LUO, H. et al. Ionic composition and transport mechanisms in microbial desalination cells. **Journal of membrane science**, v. 409, p. 16-23, 20

MADIGAN, M. T. et al. Microbiology Brock-14th Edition. Artmed Editora, 2016.

MAHMOOD, et al. Isolation and characterization of Pseudomonas stutzeri QZ1 from an anoxic sulfide - oxidizing bioreactor. Anaerob, 15, 108-115, 2009.

MEHANNA, M.; LOGAN, B. E. Microbial desalination cell for simultaneous water desalination and energy production. **Env. Sci Technol**, v. 44, p. 1-10, 2010.

MURRAY, P. R.; BARON, E. J.; JORGENSEN, J. H.; PFALLER, M. A.; YOLKEN, R. H. **Manual of clinical microbiology**. 8th ed., ASM Press, Washington DC, 2003. 2113p.

NATIONAL RESEARCH COUNCIL et al. **Committee on Advancing Desalination Technology and National Academies Press** (US) 2008 Desalination: A National Perspective vol xiv.

OLIVE TREE, A. N.; OLIVE TREE, L. A.; ANDRADE, J. S.; JUNIOR, C. Extracellular Hydrolytic Enzymes from Rizobic isolates native to central Amazonia, Amazonas, Brazil1. Food Science and Technology, Campinas, v. 26, n. 4, p. 853-860, Oct./Dec, 2006.

ONU - UNITED NATIONS ORGANISATION. **World Report United Nations Water Resources Development 2018**: Nature-based Solutions for Water Management. Available in: $\leq$ http://portalods.com.br/wp-content/uploads/2018/03/261594por.pdf> Access in: July 2019.

OTTO, M. Virulencefactorsofthe staphylococci coagulase-negative. **Front Biosci**, v. 9, n. 1, p. 841-863, 2004.

PEREIRA, C. R.; SOARES-GOMES, A. 2ª ed. **Marine Biology. Rio de Janeiro: Interciência**, 2009.

PILLING, S. **Extremophiles (types, properties, zone of extreme habitability).** Astrobiology. São José dos Campos, 2015.

PING, Q. et al. Long-term investigation of fouling of cation and anion exchange membranes in microbial desalination cells. **Desalination**, v. 325, p. 48-55, 2013.

PROAL, A. Understanding biofilms. **Bacteriality exploring chronic disease**. 2008.

QU, Y. et al. Simultaneous water desalination and electricity generation in a microbial desalination cell with electrolyte recirculation for pH control. **Bioresource technology**, v. 106, p. 89-94, 2012.

QU, Y. et al. Salt removal using multiple microbial desalination cells under continuous flow conditions. **Desalination**, v. 317, p. 17-22, 2013.

RALUY, G.; SERRA, L.; UCHE, J. Life cycle assessment of MSF, MED and RO desalination technologies. **Energy**, v. 31, n. 13, p. 2361-2372, 2006.

SCHMIEGELOW, J.M.M. The blue planet: an introduction to marine science. **Rio de Janeiro: Interciência**, 2004.

SECTION, I.; PROGRAMME POLES, Single Paragraph Os. **CONSOLIDATION ORDINANCE NO. 5, OF SEPTEMBER 28, 2017. 2001.**

SEKAR, R. et al. Early stages of biofilm succession in a lentic freshwater environment. In: **Asian Pacific Phycology in the 21st Century: Prospects and Challenges**. Springer, Dordrecht, 2004. p. 97-108.

SEMIAT, R. Energy issues in desalination processes. **Environmental science & technology, v.** 42, n. 22, p. 8193-8201, 2008.

SHANNON, M. A. et al. Science and technology for water purification in the coming decades. In: **Nanoscience and technology: a collection of reviews from nature Journals**. 2010. p. 337-346.

SILVA, C. A. R. e. **Chemical Oceanography.** Rio de Janeiro: Interciência, 2011.

SPOLIDORIO, D. et al. Microbial Morphology. **Microbiology and General and Dental Immunology**, 2015.

TORRES, C. Improving microbial fuel cells. **Membrane Technology**, v. 2012, n. 8, p. 8-9, 2012.

WALTER, X. A; GREENMAN, J.; IEROPOULOS, I. A. Oxygenic phototrophic biofilms for improved cathode performance in microbial fuel cells. **Algal research**, v. 2, n. 3, p. 183-187, 2013.

WANG, H.; REN, Z. J. A comprehensive review of microbial electrochemical systems as a platform technology. **Biotechnology advances**, v. 31, n. 8, p. 1796-1807, 2013.

Printed by Books on Demand GmbH, Norderstedt / Germany